Wireless Networks

Series editor

Xuemin (Sherman) Shen
University of Waterloo, Waterloo, Ontario, Canada

More information about this series at http://www.springer.com/series/14180

Yanru Zhang · Zhu Han

Contract Theory for Wireless Networks

 Springer

Yanru Zhang
Department of Electrical and Computer
    Engineering
University of Houston
Houston, TX
USA

Zhu Han
Department of Electrical and Computer
    Engineering
University of Houston
Houston, TX
USA

ISSN 2366-1186   ISSN 2366-1445   (electronic)
Wireless Networks
ISBN 978-3-319-53287-5   ISBN 978-3-319-53288-2   (eBook)
DOI 10.1007/978-3-319-53288-2

Library of Congress Control Number: 2017930794

Printed on acid-free paper

This Springer imprint is published by Springer Nature
The registered company is Springer International Publishing AG
The registered company address is: Gewerbestrasse 11, 6330 Cham, Switzerland

Preface

With the dramatic attention drawn by the problems of network economics in wireless communications and networking research community, many economic theories have been applied. Being rewarded with the 2016 and 2014 Nobel Prize of economic science, contract theory has been introduced to this area after the wide researches in game theory, auction theory, and pricing strategy. In this book, we are intend to provide a pure mathematical and economics-oriented approach for introducing contract theoretical concepts, in which we integrate the notions from contract theory and wireless engineering, while emphasizing on how contract theory can be applied in wireless networks. Furthermore, we plan to describe the details and challenges of modeling, analyzing, and designing contract theoretical approaches for communication and networking problems. We believe engineers and researchers in the wireless communication community who are interested in the state-of-the-art research on incentive mechanism and pricing schemes design, resource sharing and trading, cooperation, and networking for a wide range of wireless communication applications will find it useful.

Houston, TX, USA
November 2016

Yanru Zhang
Zhu Han

Acknowledgements

Thanks to all the collaborators who have also contributed to this book. They are Dr. Lingyang Song, Dr. Zaher Dawy, Dr. Walid Saad, Dr.Miao Pan, Dr. Dusit Niyato, Dr. Nguyen H. Tran, and Dr. Chunxiao Jiang. Thanks to NSF for supporting this research.

Contents

Chapter 1
Background Introduction

1.1 Introduction

Owing to the wide adoption of smart devices and fast development of the Internet, various applications and services have been introduced to bring convenience to every aspect of our daily lives; at the same time, this has brought great changes and new challenges to the design and operation of wireless networks. First, the introduction of resource demanding mobile services such as Facebook and YouTube has exponentially raised the desire for wireless access (Sesia et al. 2009). Moreover, the embedding of advanced sensors in mobile devices has led to the dramatic growth of a wide range of location-based services.

On the one hand, one can deal with the network capacity crunch by utilizing various forms of cooperation in heterogeneous wireless networking scenarios. Technologies such as device-to-device (D2D) communications, cognitive radio (CR), and small cells are being developed to off-load the cellular traffic and increase the energy and spectrum efficiency. On the other hand, fast develop in the Internet of things increases the need for cooperations from different aspects, as well. For example, an attractive solution for location-based data crunch is to do mobile crowdsourcing, in which a large group of users (with sensors embedded smart devices) cooperate to collect and transmit data regularly for the service provider. Other forms of cooperations can be found in mobile cloud computing, virtualized network, etc. In both wireless networks and Internet, it is necessary to ensure the cooperation from third parties, e.g., D2D devices, small cells, and users.

However, there lies a conflict when participating in such activities, as third parties do consume their resources, such as battery capacity and computing power (Zhao et al. 2014). Such a conflict results in reluctance from third parties to participate, which is a major impediment to the development of practically attractive traffic offloading and mobile crowdsourcing solutions. Therefore, to successfully achieve the benefits, there is a need to develop effective incentive mechanism designs for wireless networks, in order to incentivize third-party participation and improve overall operational quality.

© Springer International Publishing AG 2017
Y. Zhang and Z. Han, *Contract Theory for Wireless Networks*,
Wireless Networks, DOI 10.1007/978-3-319-53288-2_1

Contract theory is widely used in real-world economics with asymmetric information to design contracts between employer/seller(s) and employee/buyer(s) by introducing cooperation (Bolton and Dewatripont 2004). The information asymmetry usually refers to the fact that the employer/seller(s) does not know exactly the characteristics of the employee/buyer(s). By using contract theory-based models, the employer/seller(s) can overcome this asymmetric information and efficiently incentivize its employee/buyer(s) by offering a contract which includes a given performance/item and a corresponding reward/price.

Contract theory was first introduced in 1960s and has been well developed over decades. With the property of been practical in real-world economics, contract theory has been successfully applied in industrial economics and public economics, such as banking, telecommunications, and agriculture. The importance of contract theory has been well recognized recently. In 2014, French economist Jean Tirole was awarded with the Nobel Prize in Economic Sciences for he has developed deep analytical results about the essential nature of imperfect competition and contracting under asymmetric information. Two years later, Nobel Prize in Economic Sciences 2016 was awarded jointly to Oliver Hart and Bengt Holmström "for their contributions to contract theory."

Due to the properties such as inducing cooperation and dealing with asymmetric information, we envision that there is a great potential to utilize concepts from contract theory to ensure cooperation and assist in the design of incentive mechanisms in wireless networks. In wireless networks, the employer/seller(s) and employee/buyer(s) can be of different roles depending on the scenario under consideration. An employer/seller(s) can be a base station (BS), service provider (SP), and authorized spectrum owner. An employee/buyer(s) can be a small cell, smart device, user, or some other third party that is not part of the current traditional cellular network architecture. Well-designed contracts provide incentives for the contracting parties to exploit the prospective gains from cooperation. The adoption of contract theory for incentive mechanism design in future wireless networks is illustrated in Fig. 1.1.

In this book, we mainly focus our research on how to provide the necessary incentives to motivate third parties' participation in those newly introduced wireless networks and Internet, such as heterogeneous network, mobile crowdsourcing, cloud computing, and virtualized network. We are going to use contract theory to formulate the incentive mechanisms for wireless networks. For each class of the typical contract models, we provide the basic concepts, classification, and models in Sect. 1.2, as well as comparisons with other economical methods. Beyond providing a self-contained survey on classical contract theory concepts, we will further study in Sect. 1.3 the design of incentive mechanisms, especially the reward design in a contract. We then emphasize both analytical techniques and novel application scenarios in Sect. 1.4. Finally, we give summary of this chapter and the main organization of this book in Sect. 1.5.

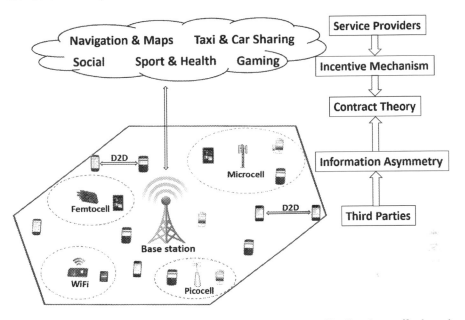

Fig. 1.1 General model for cooperation in wireless networks: (1) offloading data traffic through heterogeneous networks (small cell, cognitive radio, and D2D communication) and (2) uploading location-based data through mobile crowdsourcing

1.2 Contract Theory: Fundamentals and Classification

1.2.1 Basic Contract Concepts

Contract theory has been highly successful and active research area in economics, finance, management, and corporate law for decades. Contract theory allows studying the interaction between employer(s) and employee(s). The performance of employees tends to be better when they work harder, and the probability of a bad performance will be lower if employees place more dedication or focus on the work. By contrast, on the other hand, if an employee's compensation is independent of its performance, the employee will be less likely to put efforts into the work (Bolton and Dewatripont 2004). The design of incentive mechanism plays an important role in addressing the problem of employee incentives.

In contract theory, the solution we need to obtain is a menu of contract for employee, and the object is maximizing the employer's payoff or utility. In most cases, the problem is usually formulated as maximizing an objective function which represents the employer's payoff, subject to the *incentive compatibility* constraint that the employee's expected payoff is maximized when signing the contract and the *individual rationality* constraint that the employee's payoff under this contract is larger than or equal to its reservation payoff when not participating.

1.2.2   Classification

1.2.2.1   Adverse Selection

The *adverse selection* problem and the information about some relevant characteristics of the employees, such as their distaste for certain tasks and their level of competence/productivity, are hidden from the employer. One of the most common problems in *adverse selection* is the *screening problem*, in which the contract is offered by the uninformed party, i.e., the employer. The uninformed party typically responds to *adverse selection* by the revelation principle which forces the informed party to select contract that fits its true status. When a contract is offered by the uninformed party, it is called the *screening problem*. Otherwise, if the contract is offered by the informed party, it is called the *signaling problem*.

In the *screening problem*, it is the employer who makes the contract offer and tries to screen the information hold by the employee. Based on the revelation principle, the employer can offer multiple employment contracts (q, r) destined to different skill-level employees, where q is the employee's outcome wanted by the employer, and r is the reward paid to the employee by the employer if the given target is achieved. The outcome can be a duration of work time, a required performance, or some other outcomes that the employer wants from the employee.

In the *signaling problem*, the employee attempts to signal the employer its capability through the type of action it takes, which consists of the part of the contract. According to the revelation principle, the employee can make a contract offer as (a, r), where a is the action or effort taken by the employee before entering the labor market, which serves as a signal to reveal its productivity, and r is the reward the employee wants from the employer. The action or effort can be the employee's education level, past working experiences, or even financial status.

Most of the *adverse selection* models have the following system model. Assume there are n different types, $\theta_i, i \in \{1, \dots, n\}$, of employees, where the type represents their level of capability, competence, or etc. There exists an information asymmetry that the employer does not know the exact type of employee, but only the probability λ_i of facing a type θ_i employee.

The employer has the expected utility function

$$U = \sum_{i=1}^{n} \lambda_i \, (q_i - r_i), \tag{1.1}$$

which is the employer's received outcome from the employee minus the reward it has to pay. The employee has the utility function

$$V = \theta_i v(r_i) - q_i, \tag{1.2}$$

which is the employee's evaluation toward the reward received from the employer minus the cost of the outcome.

The problem of *adverse selection* is usually formulated as follows.

$$\max_{(q,r)} \sum_{i=1}^{n} \lambda_i \left(q_i - cr_i \right), \tag{1.3}$$

s.t.

$$(IR)\ \theta_i v(r_i) - q_i \geq 0,$$
$$(IC)\ \theta_i v(r_i) - q_i \geq \theta_i v(r_j) - q_j,$$
$$i, j \in \{1, \ldots, n\}, \quad i \neq j.$$

The employer's utility is maximized, under two constraints: individual rationality (IR) and incentive compatible (IC). The IR constraints mean the contract must ensure a nonnegative utility for all types of employees. The IC constraint guarantees the employee can only receive the highest utility when selecting the contract designed specifically for their own types. The model presented here is the discrete-type case which is also the most commonly seen one. The model can also be extended to the continuous type case which can fit into more general cases.

1.2.2.2 Moral Hazard

The problem of *moral hazard* refers to situations where the employee's actions are hidden from the employer: whether they work or not, how hard they work, how careful they are. In contrast to *adverse selection*, the informational asymmetries in *moral hazard* arise after the contract has been signed. In *moral hazard*, the contract is a menu of action–reward bundle (a, r), where a is the action or effort exerted by the employee after being hired, and r is the reward paid to the employee by the employer.

The basic model of *moral hazard* problem is as follows. The employer offers a compensation package r to an employee, which is a combination of a fixed salary t and a performance-related bonus s. The employee's performance q can be defined according to the application. During work time, the employee's effort can be regarded as taking an action a, while there is asymmetric information that the effort a is hidden from the employer who can only observe the performance q. Due to some measurement errors, the performance q is slightly different from the actual effort exerted by the user. Therefore, the performance of the user is a noisy signal of its effort. Thus, we assume that the performance q to be normally distributed with mean a and variance σ_q^2:

$$q = a + \varepsilon_q, \tag{1.4}$$

where $\varepsilon_q \sim N(\mu_q, \sigma_q^2)$, and μ is the mean. One simple form for the bonus is the linear form. By restricting the compensation package offered by the principal in the linear form, the compensation package r the employee receives from working for the employer can be written as

$$r = t + sq, \tag{1.5}$$

where t denotes the fixed compensation salary independent of performance, and s is the fraction of reward related to the employee's performance q.

The employee is usually assumed to have a constant absolute risk averse (CARA) risk preferences, which means the employee has a constant attitude toward risk as its income increases. Thus, employee's utility is represented by a negative exponential utility form:

$$u(w, a) = -e^{-\eta[r - \psi(a)]}, \tag{1.6}$$

where $\eta > 0$ is the employee's coefficient of absolute risk aversion ($\eta = -u''/u'$). A larger value of $\eta > 0$ means less incentive for the employee to implement an effort. $\psi(a)$ is the incurring cost in providing the effort a for the employer. The cost function can be assumed to be quadratic, or others according to different applications,

$$\psi(a) = \frac{1}{2}ca^2. \tag{1.7}$$

The utility of the employer is the evaluation of the working outcome q minus the compensation package r to the employee,

$$V(r, a) = E(q - r), \tag{1.8}$$

where $E(\cdot)$ is the evaluation function follows $E(0) = 0$, $E'(\cdot) > 0$, and $E''(\cdot) \geq 0$ if the principal is assumed to be risk neutral, i.e., $E''(\cdot) = 0$. Thus, the utility of the employer can be simplified as

$$U(r, a) = q - r = (1 - s)a - t. \tag{1.9}$$

The problem of *moral hazard* is usually formulated as follows.

$$\max_{a,t,s} U(r, a), \tag{1.10}$$

$$s.t. \ (IC) \ a^* \in \arg\max_a V(r, a),$$

$$(IR) \ V(r, a) \geq V(\bar{r}).$$

The employer's utility is also maximized under the IC and IR constraints. The IR constraint means the contract must ensure the employee receives a higher utility than when not participated in. The IC constraint guarantees the employee can maximize its own utility when selecting the right amount of effort.

1.2.2.3 Mixed

In practice, it is usually hard to decide which of the two problems is more important, i.e., to figure out whether it is a *moral hazard* problem or adverse selection problem. Indeed, most incentive problems are the combinations of *moral hazard* and *adverse selection*.

1.2.3 Models

1.2.3.1 Bilateral or Multilateral

Bilateral contracting is the basic one-to-one contracting model, in which there are one employer and one employee trading with each other for goods or services. However, in the multilateral case, it is usually an one-to-many contracting scenario, in which there is one employer trading with multiple employees. Despite the increased number of participants in the multilateral contracting than in the bilateral one, the interactions among the employees/buyers, such as competition and cooperation, make the multilateral contracting model more complex and show the potential of solving more sophisticated problems.

1.2.3.2 One-Dimension or Multi-dimension

Only one characteristic or task is considered in the one-dimensional contracting model. For example, the employer evaluates only one capability of the employee in the one-dimensional adverse selection model, and there is only one task assigned by the employer to the employee in the one-dimensional moral hazard model. In contrast, the employer evaluates multi-dimensional characteristics of the employee or assigns multiple tasks to the employee in the multi-dimensional contracting scenario. For example, the action a in the one-dimensional *moral hazard* model can be extended to $\mathbf{a} = (a_1, \ldots, a_n), n \geq 2$. Meanwhile, the observed performance becomes $\mathbf{q} = (q_1, \ldots, q_n)$, as well as the bonus $\mathbf{s} = (s_1, \ldots, s_n)$. As the extension of one-dimensional contracting, multi-dimensional contracting model can also be analyzed by adapting the similar methods for one-dimensional ones.

1.2.3.3 Static or Repeated

Static contracting refers to the one-shot trading between the two parties, in which the employer usually offers a take-it or leave-it contract, and the employee(s) choose to accept or reject it. Every signing of a contract will be regarded as a new one; i.e., previous trading histories will not affect the signing of the next one. While trading histories affect the next contract in the repeated contracting scenario, repeated contracting needs to solve the issues that arise with the design and renegotiation of long-term employment contracts, due to the inability of contracting parties to commit to or enforce long-term contractual agreements. Repeated iteration between contracting parties opens up new incentive issues and thus increases the complexity than in the static contracting.

1.2.3.4 Complete or Incomplete

In complete contract, legal consequences of every possible state of the world are specified at the stage of signing the contract. In reality, there might be situations when the trading parties are unable to write complete contracts at the stage of signing contract, because either it is impossibly complex for the parties to fulfill an agreement to make their contract complete or it is too costly to do so. Such an inability to describe future uncertainty is a binding constraint in designing the contract.

1.2.4 Comparisons

1.2.4.1 Market Equilibrium

In the market equilibrium, participants pay their own strategy in regard to the other's actions in each iteration and then finally reach the equilibrium. While similar to the repeated long-term contracting scenario in contract theory, participants dynamically change their strategy as if they are playing a game. After repeated interactions and renegotiations, both parties can reach an agreement. Thus, we see that the market equilibrium is the repeated contracting case in contract theory, and different scenarios can fit either into the problem of *adverse selection* or *moral hazard*.

1.2.4.2 Auction Theory

In auction theory, there is one seller with an item to sell and multiple bidders with reservation prices competing for it. Meanwhile, in the multilateral *adverse selection*, there are one seller and multiple buyers with their own private information which is the same case as the bidder's reservation prices during the auction. Thus, we see that auction theory is the multilateral *adverse selection* contracting problem in contract theory.

1.2.4.3 Pricing Strategy

The problems that pricing strategy and contract theory can solve have some overlaps. They two are similar to each other in the sense that they can adjust the price/reward to sell a product or service at the seller/employer's maximal profitability. However, pricing strategy and contract theory's major focuses differ from each other, since pricing strategy mainly focuses on the relation between pricing and marketing, which can be used to beat the business competitors. While contract theory places the emphasis on studying the interactions between employers and employees, which is helpful for regulators to design incentive mechanisms.

From the three economic models, we can see, first, market equilibrium and auction theory, as the special cases of contract theory, have already been widely studied. Second, pricing strategy shows a different research direction as contract theory. To design efficient incentive mechanism, contract theory seems to be an excellent approach and has many unexplored areas to reveal.

1.3 Contract Theory: Reward Design

In contract theory, the objective is to motivate employees by offering a reward, in trading with a level/quality of service, outcome, performance, or target. Thus, we see that the reward determines whether the employee can be fully motivated by the incentive mechanism. Given the large number of models in contract theory, the reward design varies in different contracting scenarios. The design and classification of reward are illustrated in Fig. 1.2 and will be discussed in detail in this section.

1.3.1 Dimension of Rewards

From Sect. 1.2.3.2, we know that there can be one- or multi-dimensional contract-theoretic models, depending on how many aspects of capability do the employer evaluate the employee, or how many tasks does the employer assign the employee. Most existing literature on incentive mechanism design in wireless networks adopts the one-dimensional reward model. One example is the reimbursing scheme proposed by Gao et al. (2014b), which is an usage-based reward design to motivate subscribers to operate as mobile Wi-fi hot spots to provide Internet connectivity for others.

One-dimensional model becomes inefficient when employees are required to have multiple capabilities or supposed to work on several tasks. First, the employee's action set becomes richer than what the one-dimensional model has described. Second, there is a risk that one-dimensional reward will induce employees to overwhelmingly focus on the part that will be rewarded and to neglect the other components. Taking Yelp for an example, which is a popular mobile crowdsourcing app in North American used to locate and review restaurants/bars, Yelp users who act as employees do not only make location-based check-ins, upload photographs, and write reviews of the restaurants and bars, but they are also encouraged to invite new friends to sign up. For example, if Yelp only rewards users on the number of reviews, the quality of a review such as length, correctness, and objectiveness will not be considered.

Fig. 1.2 Designing of reward in a contract

Given different aspects of capability or multiple tasks to evaluate, by assigning different weights of rewards in multiple dimensions, the employer can drive employee's incentive on perusing certain capabilities or tasks, which can affect the employer's utility, in return. One current application of multi-dimensional reward is from Karma (2012), where Karma is an Internet service provider based in the USA. Karma provides 100 MB to new guest users for free and reward users who bring in more users by wirelessly advertising the service.

Thus, in certain scenarios, one-dimensional reward needs to be modified into multi-dimensional ones, so that the employer can drive employee's incentives by assigning different reward weights on different tasks. Regardless of the dimension that the reward design chooses, a qualified mechanism must reward employee's effort in a comprehensive way. On the one hand, for simple cases where an one-dimensional reward is sufficient to drive employee's motivation, a multi-dimensional reward mechanism costs extra effort and resource to design. On the other hand, for complicated tasks, the reward design must be adjusted to multi-dimension, so that employee's incentive can be well maintained and driven. In reward design, there is a trade-off between completion and efficiency, and thus, we should model the dimension of the reward according to the actual scenario under consideration.

1.3.2 Rewards on Absolute Performance or Relative Performance

The problem of how can the reward be decided in accordance with the employee's performance also needs attention. Referring to the reward designs in job markets, sports, and games, generally there are two methods one can refer to: evaluate the employee's absolute performance or the relative performance.

- *Absolute performance-related reward*: The reward is positively related with the employee's absolute performance.
- *Relative performance-related reward*: The reward is given based on the ranks that the employees achieved by listing the multiple employees' performance in an ascending or descending order.

Absolute performance-related reward is a widely accepted incentive mechanism in real economics as it captures the fundamental aspect of providing necessary and efficient incentives for employees. Piece rate, efficiency wages, and stock options are widely used forms of absolute performance reward in the job market. Despite the usage-based reward in (Gao et al. 2014b) mentioned previously, the work in (Luo et al. 2014) also derives the performance and reward-dependent function to attract a high amount of sensing data from participating users in wireless networks. Another example is (Guo et al. 2015), in which incentive mechanism has been developed to encourage the cooperation of mobile terminals (MTs) in wireless cellular networks to reduce the energy consumption of the other MTs. The MT who contributes to help will receive a price consistent with its transmitting data rate.

However, there are two disadvantages of the absolute performance-related reward. First, in order to pay less reward, the employer has a strong incentive to cheat by claiming that employees had poor performances. Second, this mechanism is vulnerable to *common shock* which is originally used to denote macroeconomic conditions such as economic boost or depression (Green and Stokey 1983). If there is a positive/negative mean that affects employees' performances at the employer's observation, then it will lead to an abnormal increase/decrease of reward in the end.

While it has been proven that the relative performance-related reward design can filter out this *common shock* problem (Green and Stokey 1983), winners receive the amount of reward based on the rank they achieved, which is easy to measure and hard to manipulate (Bolton and Dewatripont 2004). In addition, the employer has no incentive to cheat as it has to offer the fixed amount of rewards no matter who wins. Tournament is the most widely known form of reward for the relative performance, in which the one with better performance ranks higher and rewarded more. Besides, there are two other special forms of ROT: the multiple winners (MW) and winner-take-all (WTA). In the MW tournament, several top winners share the reward equally, while in the WTA tournament, the entire reward is awarded to the highest ranked user, which is a special case of MW with only one winner.

1.3.3   Reward in Bilateral or Multilateral Contracting

Despite the previous aspects, different trading scenarios also affect the design of an incentive mechanism, i.e., the reward. Next, we are going to talk about how to design reward in bilateral and multilateral contracting scenarios.

1.3.3.1   Contract with Single Employee

When the employer signs a contract with a single employee, we can design the reward by considering only the single employee's absolute performance instead of the others. Examples in wireless networks are the previously mentioned three works (Gao et al. 2014b; Luo et al. 2014; Guo et al. 2015). However, even though there is no other employee to compete with the employee, the relative performance-related reward can still be applied. One common form of the relative performance-related reward for a single employee is to set up a specific threshold and a reward of the targeted performance. If the employee's absolute performance can achieve the given threshold, a fixed reward will be given to the employee. Otherwise, the employee cannot receive the reward. In fact, we can regard it as the employee competes with the threshold.

1.3.3.2   Contract with Multi-employee

When the employer designs the contract toward multiple employees, the absolute performance-related reward still works quite well and is a widely accepted method in real economics. Furthermore, there are some other forms of absolute performance-related rewards. One widely adopted method is to group employees first and then reward employees by their aggregated performance in each group. There is a shortcoming with this incentive mechanism, i.e., there is a risk of free riding of some employees on the other employees' efforts. Usually, the absolute performance-related reward design is more commonly seen in contracting with multi-employee. The employees can compete with each other as in a tournament and have the incentives for higher rewards by performing better.

1.4   Applications in Wireless Networks

In this section, we are going to introduce several applications of contract theory models in wireless networks. To be consistent with the classification of contract theory problems in Sect. 1.2.2, the following three subsections are wireless network applications of models from *adverse selection*, *moral hazard*, and a mixed of the two, respectively.

1.4.1 Adverse Selection

The applications of bilateral, one-dimensional, and static *adverse selection* in wireless networks are the most widely seen models. This model is first used to solve the problem of spectrum sharing in cognitive radio network (CRN) by Gao et al. (2011). In this work, a primary user (PU) acts as an employer who sets the spectrum trading contract as *(qualities, prices)*, and the second users (SUs) act as an employee to choose which one for purchasing. Another application for CRNs can be found in Gao et al. (2014a), in which the authors also model the PU and SUs as the employer and employees, respectively. Then, design the *(performance, reward)* in contract as *(relaying power, spectrum accessing time)*.

With the same model, a different application area is by Duan et al. (2012) in designing incentive mechanisms for smartphone users' collaboration on both in data acquisition and in distributed computing. The SP acts as an employer, and smartphone users will be employees. Rewards will be paid according to the amount of data users have collected and the distributed computing load users have taken. In the OFDM-based cooperative communication system, Hasan and Bhargava (2013) uses contract theory to tackle the source node's relay selection problem. The offers/contracts consist of a menu of desired signal-to-noise ratios (SNRs) at the destination and corresponding payments. In Chap. 2, we will apply the *adverse selection* model in cellular traffic offloading through D2D communication, by offering rewards to encourage content owners to participate and cooperate with other devices via D2D. We will model the BS as employer and D2D user as employee and solve contract bundle with a required performance and an absolute performance-related reward. The performance is defined as a certain data rate that the UE must provide during the D2D communication.

1.4.2 Moral Hazard

Compared to the wide adoption of the *adverse selection* problem, the *moral hazard* problem has hardly been applied in wireless networks by now. However, having seen a great potential of this model, we have done some preliminary applications of moral hazard in the area of mobile crowdsourcing. As mentioned in the beginning of this chapter, many users hesitate to participate in mobile crowdsourcing with certain concerns, which results in a serious impediment to the exploitation of location-based services. By adopting the *moral hazard*, the incentive mechanism can be designed by regarding the SP "employs" users to upload location-based data and reward them by their performance. Thus, the first application we are going to talk about in Chap. 3 is a basic *moral hazard* model which only considers a single user and single lateral in mobile crowdsourcing. Then, in the next Chap. 4, we consider the general case of a large group of users as employees, and thus, the multilateral *moral hazard* model is applied. In particular, we consider the mobile users competing in the crowdsourcing to win reward as in a tournament, and they are rewarded by their rank orders, i.e., relative performance in the mobile sourcing activity.

Besides mobile crowdsourcing, we have also studied the area of mobile cloud computing and moved forward to the multi-dimensional model for further performance improvement. In particular, we focus on the newly introduced concept of fog computing which aims at providing time-sensitive data services with low latency, location awareness to end users. One key feature of this model is the designing of a payment plan from the network operator (NO) to fog nodes (FNs) for the rental of their computing resources, such as computation capacity, spectrum, and transmission power. To better solve the problem of how to design the efficient payment plan to maximize the NO's revenue while maintaining FN's incentive to cooperate, we propose a multi-dimensional moral hazard model which considers the FNs' characteristics such as location, computation capacity, storage, and transmission bandwidth in Chap. 5.

1.4.3   Mixed Problem

Given the applications of the two basic problems: *adverse selection* and *moral hazard*, we can proceed to the mixed problem in wireless networks when both of the two were present. The mix problem can also be found in spectrum trading between the PU and SU in CRNs, or infrastructure provider (InP) and SP in virtualized wireless networks. The problem of *adverse selection* arises since the PU/InP may not be fully aware of the SU/SP's capability in utilizing the spectrum to generate revenue, i.e., what is the SU/SP's probability of successfully making a profit from the service it provides. Moreover, there is a problem of *moral hazard* as the PU/InP neither knows how much effort the SU/SP will put into running its "business." Thus, spectrum trading that involves both *adverse selection* and *moral hazard* can be solved by designing a financing contract, as when we buy a car or a house. The main problem that needs to solve is how to design the down payment and installment payment in the financing contract, and the detail work can be found in Chap. 6.

1.4.3.1   Incomplete Contracts in Wireless Networks

The previous problems described are all solved by complete contracts, while in wireless networks, there are also problems that are in need of designing incomplete contracts to solve. One typical example of incomplete contract in economy is relationship-specific investments. In virtualized wireless network, the complementary relationship between infrastructure provider (InP) and service provider (SP) is usually a long-term supply contract, and details of trades are left to be specified in the future. Thus, the returns of the InP and SP depend on their bargaining positions, ex post, and investments, ex ante. As a result, the InP and SP may hesitate to have specific investment, since it may put them at risk of no return. The problems of determining how the ownership of the resources affects the InP and SP's incentives to invest and how to choose the most efficient investments in an MVN are studied in Chap. 7.

1.5 Summary

In this chapter, we have provided the fundamental concepts of contract theory and introduced the potential applications for each class of the typical contract problems: *adverse selection* and *moral hazard*. Specially, we have investigated the design of reward, which is the most critical element in an incentive mechanism design. We have also provided a detailed description on how to use such contract-theoretic tools in several wireless applications, such as spectrum trading in cognitive radio network, D2D communication, mobile crowdsourcing, fog computing, and virtualized network. From those works, we have seen contract theory as a useful framework to design incentive mechanisms to motivate the third party's participation in emerging wireless networks. In the end, the future research directions for contract theory applications in wireless networks are discussed, including the promising areas and potential techniques from contract theory. In a nutshell, this chapter is expected to provide an accessible and holistic introduction on the use of new techniques from contract theory to address incentive problems in emerging wireless networks.

The rest of the book will go further into each application for detail modeling and analysis. The organization of this book is as follows. The first application in Chap. 2 will be the *adverse selection* problem in D2D communication. Then, the *moral hazard* problem in mobile crowdsourcing and fog computing will be described in Chaps. 3, 4, and 5, with each chapter provides the single-user single-reward model, multi-user single-reward model, and single-user multi-reward model, respectively. After discussing the applications of the two basic problems, we proceed to the mixed problem of spectrum trading in cognitive radio network in Chap. 6. After discussing those complete contract cases, we will give one application of incomplete contract in Chap. 7 which is aiming at giving efficient investments in a mobile virtualized network. Finally, conclusions and some possible future works are mentioned in Chap. 8.

References

Bolton P, Dewatripont M (2004) Contract theory. The MIT Press, Cambridge, MA

Duan L, Kubo T, Sugiyama K, Huang J, Hasegawa T, Walrand J (2012) Incentive mechanisms for smartphone collaboration in data acquisition and distributed computing. In: IEEE Proceedings of INFOCOM, Orlando, FL

Gao L, Wang X, Xu Y, Zhang Q (2011) Spectrum trading in cognitive radio networks: A contract-theoretic modeling approach. IEEE J Select Areas Commun 29(4):843–855

Gao L, Huang J, Chen Y, Shou B (2014a) Contract-based cooperative spectrum sharing. IEEE Trans Mob Comput 13(1):174–187

Gao L, Iosifidis G, Huang J, Tassiulas L (2014b) Hybrid data pricing for network-assisted user-provided connectivity. In: INFOCOM, Proceedings IEEE, Toronto, Canada

Green JR, Stokey NL (1983) A comparison of tournaments and contracts. J Polit Econ 91(3):349–364

Guo Y, Duan L, Zhang R (2015) Optimal pricing and load sharing for energy saving with communications cooperation. IEEE Trans Wirel Commun, pp: 951–964

Hasan Z, Bhargava V (2013) Relay selection for ofdm wireless systems under asymmetric information: A contract-theory based approach. IEEE Trans Wirel Commun 12(8): 3824–3837

karma (2012) Meet karma. Technic report. https://yourkarma.com/

Luo T, Tan H, Xia L (2014) Profit-maximizing incentive for participatory sensing. In: IEEE Proceedings of INFOCOM, Toronto, Canada

Sesia S, Toufik I, Baker M (2009) LTE: The UMTS long term evolution. Wiley, New York

Zhao D, Li X, Ma H (2014) How to crowdsource tasks truthfully without sacrificing utility: online incentive mechanisms with budget constraint. In: IEEE Proceedings of INFOCOM, Toronto, Canada

Chapter 2
Incentive Mechanisms for Device-to-Device Communications in Cellular Networks with Adverse Selection

2.1 Introduction

The proliferation of highly capable mobile devices, such as smartphones and tablets, coupled with the introduction of resource demanding mobile services has exponentially increased the demand for wireless access Sesia et al. (2009). A tremendous amount of mobile data, especially mobile video traffic, is rapidly straining the capacity of current wireless cellular networks (Cisco 2011). Consequently, novel wireless networking paradigm is needed to meet the challenges of this unprecedented growth in the demand for the wireless spectrum (Camps-Mur et al. 2013).

To deal with this wireless capacity crunch, device-to-device (D2D) communication underlaid over cellular network has recently been proposed as a means to boost the overall wireless network capacity (Xu et al. 2014). D2D communication benefits from the fact that two user equipments (UEs) in proximity of one another can establish a direct communication link over the licensed band while bypassing the cellular infrastructure such as the base stations (BSs). One common form of D2D communication is the network-controlled one in which the BS manages the switching between direct and cellular links (Min et al. 2011a). Due to the proximity of the involved users, if well-designed, D2D communication can dramatically improve the wireless network capacity while reducing energy consumption (Quek et al. 2013). It can also assist in off-loading the cellular traffic from the BSs while extending their coverage (Yaacoub 2014).

If UEs' resource blocks (RBs) can be shared, local users will be able to exchange data (Song et al. 2014). For example, the BS can send a frequently requested content to a number of devices who, in turn, can utilize D2D communication to spread the content to other interested users (Qualcomm 2012). By doing so, within a certain geographical area, instead of servicing a request multiple times, the BS would only transmit contents which are not locally available. In this case, the BS's traffic is significantly reduced, and thus, the cellular network capacity is increased. To successfully achieve this goal, one main design challenge is to incentivize content owners to participate and cooperate with other devices via D2D. If most users are unwilling to

© Springer International Publishing AG 2017
Y. Zhang and Z. Han, *Contract Theory for Wireless Networks*,
Wireless Networks, DOI 10.1007/978-3-319-53288-2_2

provide their contents via D2D communication, then the BS will still need to serve the users via the conventional cellular network. Consequently, it is unable to increase the network capacity. Clearly, the willingness of users to participate and share data is of great importance to reap the benefits of D2D over cellular in terms of improved capacity and traffic off-load.

Indeed, it is necessary to introduce effective incentive mechanisms that can encourage users to participate in content sharing. In order to provide incentives, the BS can offer rewards to users' UEs for the usage of their resources (storage, power, time, etc.) as well as for potential privacy risks arising from D2D, since UEs' RBs are open to the BS. For example, if the user is willing to share its content and assists the BS to transmit the data, the BS will offer a reward to compensate for this user's participation. The reward can be in the form of monetary remuneration or free data among others (Golrezaei et al. 2013).

Intuitively, a well-designed incentive mechanism should reward UEs based on their contributions: devices that contribute more must get higher rewards than devices with less contributions. Users with high preference toward participation will be more likely to contribute. However, each user will attempt to harness as much reward as possible by claiming that it is a high preference user, which brings difficulty to the BS in reward design. This problem is exacerbated by *information asymmetry*—the BSs may not be aware of the actual preference, which is naturally known by the users. To this end, our main goal is to propose an incentive mechanism by overcoming this information asymmetry in a D2D network as shown in Fig. 2.1.

In this respect, there is a need to design a mechanism in which UEs will be rewarded in accordance with their preference. Contract theory, a powerful framework

Fig. 2.1 The reward assignment problem faced by the BS

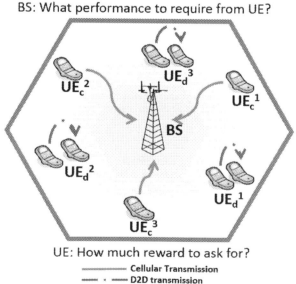

BS: What performance to require from UE?

UE: How much reward to ask for?

from microeconomics, provides a useful set of tools for modeling incentive mechanisms under information asymmetry (Werin and Wijkander 1992). Using contract theory, one can analyze the interactions between an employer who is trying to offer proper contracts and employees whose skills are not known a priori (Bolton and Dewatripont 2004). A contract is essentially a certain reward that will be given to the employee in return for its services. In a D2D context, this contractual situation can be used to study the interactions between BSs, acting as employers and, UEs, acting as devices whose preferences are unknown to the BSs. Here, the contract will represent the rewards provided by the BS to a certain D2D-capable UE who will provide the required resources and quality-of-service via D2D participation. The main advantages of adopting contract theory in a D2D scenario include the following: (1) ability to incorporate semi-distributed network control in which the BS can control the D2D communication links; (2) notions such as self-revealing contracts suitable to handle information asymmetry, and (3) ability to devise optimal reward and incentive mechanisms that can induce cooperation between UEs.

The main contribution of this chapter is to leverage the use of contract theory for introducing D2D incentive mechanisms under information asymmetry. In particular, we view the D2D sharing problem, as a contract-theoretical model in which the BS hires the UEs as employees to fulfill the content transmission task. The BS, as an employer, offers contracts to the UEs that specify different performance-reward combinations for different UE preferences. The UEs, as employees, select contracts that are the best fit to their own preferences. Under this scenario, the BS can efficiently reward the users according to their performance and thus motivate users to participate in D2D communication.

For the studied D2D contract model, we provide the necessary and sufficient conditions for contract feasibility. Here, contract feasibility implies that when users join in, they receive the reward that covers their cost and in accordance with their true preference. In addition, we study and analyze the problem under two key scenarios: the discrete (finite) type and continuum (infinite) type. To implement the proposed contract-theoretical D2D model, we propose a novel algorithm that can allow the BS and UEs to interact and then optimize the network capacity while guaranteeing a desired network quality-of-service (QoS). Simulation results show that the proposed contract-theoretical model can guarantee UEs receive positive payoffs and compatible incentives. We also study the system performance when the contract-theoretical model is implemented in a D2D-underlaid cellular network. The optimal contract gives the highest BS utility and social welfare as shown in the simulations. By varying the cellular network size, maximum D2D communication distance, and UE-type numbers, we see the physical layer parameters' impacts on the system performance.

The rest of this chapter is organized as follows. Section 2.2 provides a detailed literature survey. The system model is provided in Sect. 2.3. The optimal contract solution of discrete-type case is presented in Sect. 2.4, followed by the optimal contract solution in continuum-type scenario. The simulation results are shown in Sect. 2.5. Finally, summaries for this chapter are given in Sect. 2.6.

2.2 Related Work

D2D communication has been subject to many recent research works such as in (Min et al. 2011c; Yu et al. 2011). Due to the shared resources between direct D2D communication and traditional infrastructure-based communication, new resource allocation techniques are needed for D2D deployment (Zulhasnine et al. 2010). One major challenge in D2D is interference management (Song et al. 2014). The common mechanism is to limit maximum transmit power of D2D transmitter so as not to generate harmful interference from D2D systems to cellular networks (Min et al. 2011b).

Some interference management strategies are also proposed to enhance the overall capacity of cellular networks and D2D system. For example, the work in (Tanbourgi et al. 2014) introduces the idea of cooperative interference cancellation (CIC) between close-by UEs using D2D communications for improving the throughput of cellular networks in the downlink (DL) period. Another work in (Zhang et al. 2013) formulates the interference between different D2D and cellular communication links as an interference-aware graph and proposes an interference-aware graph-based resource-sharing algorithm. Several works study the use of D2D communication as a means to optimize resource usage and maintain an efficient coexistence between the D2D services and main cellular network (Xing et al. 2009).

Despite the large body of work on interference management and resource allocation in D2D communication, to our knowledge, few existing works have addressed to the problem of providing incentives for users to participate in cellular D2D. Moreover, using contract theory for network-controlled D2D has not been studied in existing works.

Here, we note that the contract theory has been used in areas such as mobile cloud computing and cognitive radio. For instance, in (Knapper et al. 2011), the authors study the use of contract theory as a means to optimize the economic revenues of a cloud server in a mobile cloud computing environment. Existing works such as (Gao et al. 2011, 2013), and (Duan et al. 2014a) focus on the efficiency of resource allocation in cognitive radio networks. The work in (Jin et al. 2012) introduces the concept of insurance into the model, in which if the primary owner (PO) cannot provide the channel purchased by a secondary user (SU), PO needs to pay a certain amount of indemnity to the SU. In (Gao et al. 2014), the authors develop a contract-theoretical mechanism to model the possibility of secondary users relaying data for primary users to improve data rates. The work in Duan et al. (2014b) develops the incentive compatible contracts to encourage users to participate in data acquisition and distributed computing programs.

However, potential interference caused by resource sharing makes it difficult to implement existing contract-theoretical models directly into the D2D underlaid cellular network. In summary, while resource allocation and interference management in D2D communication have been widely studied, no literature has investigated the problem of providing incentives for users to engage in D2D underlaid cellular networks using contract theory as proposed here.

2.3 System Model

Consider a cellular network with one BS, several cellular UEs, and D2D UE pairs. In each UE pair, there is one content requester (receiver) and one candidate content provider (transmitter). The UE receivers can receive data from the BS, or from their corresponding UE transmitters through D2D communications. In order to offload traffic from the network's infrastructure, the BS will offer a contract that can effectively motivate the content provider to use, when possible, D2D communication to deliver the content.

The UEs are heterogeneous with different preference toward joining D2D communication, in terms of personal favor, battery level, and storage capacity. Naturally, there is an information asymmetry between the BS and the UE. The UE is aware of its own preference while the BS may not have that information. Thus, to overcome the information asymmetry, the BS will specify a performance-reward bundle contract $(T(R), R)$, where T is the reward to the UE, R is the D2D performance required from the UE, and $T(R)$ is a strictly increasing function of R. Intuitively, better performance should be rewarded more and vice versa, which is called *incentive compatible*.

2.3.1 Transmission Data Rate

The performance R is measured by the UE's transmission data rate. We consider the uplink (UL) scenario since UL resource sharing in D2D communications only affects the BS, and the incurred interference can be mitigated by BS coordination (Feng et al. 2013).

The transmission data rate is related to the signal to interference plus noise ratio (SINR). In a cellular network with D2D underlaid, the receiver suffers interference from cellular and D2D communications due to resource sharing. When D2D communication is in the UL band, the source UE transmits data to the destination UEs using the uplink band of the cellular band. The interference comes from the other UEs (both cellular UE and D2D UE) (Xu et al. 2012). Thus, the transmission data rate of a D2D UE i in the UL band with co-channel interference is given by

$$R_i = W \log_2 \left(1 + \frac{P_i |h_{ir}|^2}{P_c |h_{cr}|^2 + \sum_{i'} P_{i'} |h_{i'r}|^2 + N_0}\right), \tag{2.1}$$

where i' is the UE with $i' \neq i$, P_c, P_i and $P_{i'}$ are the transmit powers of the cellular transmitter UE c and D2D transmitters UE i and i', respectively, h_{cr}, h_{ir} and $h_{i'r}$ are the channel gain between D2D receiver and cellular transmitter c and D2D transmitters i and i', respectively, N_0 is the additive white Gaussian noise (AWGN), W is the channel bandwidth. Hereinafter, without loss of generality, we assume that $W = 1$. $\sum_{i'} P_{i'} h_{i'r}^2$ represents the interference from the other D2D pairs that share spectrum resources with link UE pair i.

2.3.2   User Equipment Type

We define the UE type to be a representation of each UE's preference toward joining D2D communication. Given a fixed reward, a high-type UE will be more eager to contribute in the transmission and provide a high data rate. Naturally, high-type UEs are more preferred by the BS and will receive more reward. Here, we consider that the number of UE types belong to discrete, finite space. In Sect. 2.4.2.2, we will extend the results to the continuum case.

Definition 2.1 There are N D2D UE pairs in a D2D underlaid cellular network. The UEs' preferences are sorted in an ascending order and classified into N types: type-1, ..., type-i, ..., and type-N. The type of UE includes properties such as the privacy concern, battery remain, and the willingness to share data. θ_i denotes the type of UE and follows

$$\theta_1 < \cdots < \theta_i < \cdots < \theta_N, \quad i \in \{1, \cdots, N\}. \tag{2.2}$$

A higher θ implies more willingness to participate and contribute to the D2D communication. Here, we write the contract designed for *type-i* UE as (T_i, R_i). The BS does not know the type of UE; however, it has knowledge of the probability that a UE belongs to *type-i*, which is represented by λ_i, with $\sum_{i=i}^{N} \lambda_i = 1$.

Instead of offering the same contract to all UEs, the BS will offer different contract bundles according UE-type θ. The UEs are free to accept or decline any type of contracts. If the UE declines to receive any contract, we assume that the UE signs a contract of $(T(0), 0)$, where $T(0) = 0$. In the following subsections, we will give the utility function of the BS and UEs based on the signed contract.

2.3.3   Base Station Model

For a BS that employs a *type-i* UE as a D2D content provider, a proper utility function can be defined as the increased data rate by establishing a D2D communication

$$U_{\text{BS}}(i) = R_i - cT_i, \tag{2.3}$$

where $c > 0$ is the BS's unit cost, R_i is the required transmission rate UE must provide, and T_i is the reward the BS needs to pay in the contract bundle (T_i, R_i). Here, we assume that the reward to the UE is a certain amount of free data. The utility of the BS is the transmission data rate gained from D2D communication, minus the reward to UEs. For D2D communication to be beneficial for the BS, it is clear from (2.3) that we must have $R_i - cT_i \geq 0$. Otherwise, the BS will choose not to underlay D2D communication.

As there are N types of UE pairs, each with a probability λ_i, the expected utility of the BS can be represented by

$$U_{\mathrm{BS}} = \sum_{i=1}^{N} \lambda_i \left(R_i - c T_i \right). \tag{2.4}$$

2.3.4 User Equipment Model

The utility function of a *type-i* UE employed based on a contract (T_i, R_i) during D2D communication is

$$U_{\mathrm{UE}}(i) = \theta_i v(T_i) - c' R_i, \tag{2.5}$$

where $v(T_i)$ is the evaluation function regarding the rewards, which is a strictly increasing concave function of T, where $v(0) = 0$, $v'(T) > 0$, and $v''(T) < 0$ for all T, and c' is the UE's unit energy cost on providing the required transmission rate. For simplicity, we assume $c' = 1$. The utility of a UE is the received rewards minus the cost in terms of power consumption. Given the utility function in (2.5), the UE chooses the bundle that maximizes its own payoff.

2.3.5 Social Welfare

The network social welfare is the summation of the BS and UEs' utilities. As the number of D2D UE transmitters and number of UE types are all equal to N, the number of UE belongs to each type is 1. Assume that the distribution of the UE type is uniform, then summing up (2.3) and (2.5) from 1 to N, we have

$$\Pi = \sum_{i=1}^{N} [U_{\mathrm{BS}}(i) + U_{\mathrm{UE}}(i)] = \sum_{i=1}^{N} [\theta_i v(T_i) - c T_i]. \tag{2.6}$$

The transmission data rate is the internal transfer between the BS and UE and is canceled out.

2.4 Proposed Solution

In this section, we solve the BS's network capacity maximization problem. First, we will derive the necessary constraints that support the feasibility of the contract. Then, we will formulate the optimization problem and extend to the continuum-type case. Finally, we propose an algorithm for practical implementation.

2.4.1 Conditions for Contract Feasibility

To ensure that the UE has an incentive to off-load BS traffic via D2D communication, the contract that a UE selects needs to satisfy the following constraint.

Definition 2.2 Individual Rationality (IR): The contract that a UE selects should guarantee that $U_{\mathrm{UE}}(i)$ is nonnegative,

$$U_{\mathrm{UE}}(i) = \theta_i v(T_i) - R_i \geq 0, \quad i \in \{1, \cdots, N\}. \tag{2.7}$$

To motivate a UE's participation, the received reward must compensate its power consumption during D2D communication. If $U_{\mathrm{UE}}(i) < 0$, the UE will choose not to establish the D2D communication. This case can be formally captured by the case in which the UE signs the contract of $(T(0), 0)$.

If a *type-i* UE selects the contract (T_j, R_j) intended for *type-j* UE, the utility that the *type-i* UE receives is

$$U'_{\mathrm{UE}}(i) = \theta_i v(T_j) - R_j, \quad i, j \in \{1, \cdots, N\}, \quad i \neq j. \tag{2.8}$$

As we previously discussed, we want to design a contract such that *type-i* UE would prefer the (T_i, R_i) contract over all the other options. In other words, a *type-i* UE receives the maximum utility when selecting contract (T_i, R_i). The contract is thus known to be as a *self-revealing contract* if and only if the following constraint is satisfied.

Definition 2.3 Incentive Compatible (IC): UEs must prefer the contract designed specifically for their own types, i.e.,

$$\theta_i v(T_i) - R_i \geq \theta_i v(T_j) - R_j, \quad i, j \in \{1, \cdots, N\}, \quad i \neq j. \tag{2.9}$$

The IR and IC constraints are the basic conditions needed to ensure the incentive compatibility of a contract. Beyond the IR and IC constraints, there are several more conditions that must be satisfied.

Lemma 2.1 *For any feasible contract (T, R), $T_i > T_j$ if and only if $\theta_i > \theta_j$, and $T_i = T_j$ if and only if $\theta_i = \theta_j$.*

Proof We prove this lemma by using the IC constraint in (2.9). First, we prove the sufficiency: If $\theta_i > \theta_j$, then $T_i > T_j$.

According to the IC constraint, we have

$$\theta_i v(T_i) - R_i \geq \theta_i v(T_j) - R_j \quad \text{and} \tag{2.10}$$
$$\theta_j v(T_j) - R_j \geq \theta_j v(T_i) - R_i, \tag{2.11}$$

with $i, j \in \{1, \cdots, N\}$, $i \neq j$. We add the two inequalities together to get

$$\theta_i v(T_i) + \theta_j v(T_j) \geq \theta_i v(T_j) + \theta_j v(T_i), \tag{2.12}$$
$$\theta_i v(T_i) - \theta_j v(T_i) \geq \theta_i v(T_j) - \theta_j v(T_j),$$
$$v(T_i)(\theta_i - \theta_j) \geq v(T_j)(\theta_i - \theta_j).$$

As $\theta_i > \theta_j$, we must have $\theta_i - \theta_j > 0$. Divide both sides of the inequality, we have $v(T_i) > v(T_j)$. From the definition of $v(T)$, we know that v is a strictly increasing function of T. As $v(T_i) > v(T_j)$ holds, we must have $T_i > T_j$.

Next, we prove the necessity: if $T_i > T_j$, then $\theta_i > \theta_j$. Similar to the first case, we start with the IC constraint in (2.10)–(2.12). Using a similar process, we can obtain

$$\theta_i [v(T_i) - v(T_j)] \geq \theta_j [v(T_i) - v(T_j)]. \tag{2.13}$$

As $T_i > T_j > 0$ and $v(T)$ is strictly increasing with T, we must have $v(T_i) > v(T_j)$ and $v(T_i) - v(T_j) > 0$. Thus, by dividing both sides of the inequality, we get $\theta_i > \theta_j$. As a result, we have proved that $\theta_i > \theta_j$ if and only if $T_i > T_j$.

Using the same process, we can easily prove that $T_i = T_j$ if and only if $\theta_i = \theta_j$.

From Lemma 2.1, we know that if $\theta_j < \theta_i$, then $T_j < T_i$ must hold. Thus, a UE of high type should receive more reward than a UE of low type. If two UEs receive the same reward, they must belong to the same type and vice versa. Given our assumption in Definition 2.1 that $\theta_1 < \cdots < \theta_i < \cdots < \theta_N$, we have $T_1 < \cdots < T_i < \cdots < T_N$. Indeed, we can give a definition of this property.

Definition 2.4 Monotonicity: For any feasible contract (T, R), the reward T follows

$$0 \leq T_1 < \cdots < T_i < \cdots < T_N. \tag{2.14}$$

Monotonicity implies that the UEs of higher type, i.e., with higher preference toward participation. From the property in monotonicity, we can have the following proposition.

Proposition 2.1 *As a strictly increasing function of T, the contribution R satisfies the following condition intuitively*

$$0 \leq R_1 < \cdots < R_i < \cdots < R_N. \tag{2.15}$$

Proposition 2.1 shows that an incentive compatible contract requires a high performance of UE if it receives a high reward and vice versa.

Lemma 2.2 *For any feasible contract (T, R), the utility of each type of users must satisfy*

$$0 \leq U_{UE}(1) < \cdots < U_{UE}(i) < \cdots < U_{UE}(N). \tag{2.16}$$

Proof From Definition 2.4 and Proposition 2.1, we know that UEs who ask for more rewards must be able to provide larger transmitting rates, i.e., the two constraints $T_i > T_j$ and $R_i > R_j$ are imposed together. If $\theta_i > \theta_j$, we have

$$U_{\text{UE}}(i) = \theta_i v(T_i) - R_i \geq \theta_i v(T_j) - R_j \quad (IC) \tag{2.17}$$
$$> \theta_j v(T_j) - R_j = U_{\text{UE}}(j).$$

Now, we have $U_{\text{UE}}(i) > U_{\text{UE}}(j)$ when $\theta_i > \theta_j$. As $\theta_1 < \cdots < \theta_i < \cdots < \theta_N$, then $0 \leq U_{\text{UE}}(1) < \cdots < U_{\text{UE}}(i) < \cdots < U_{\text{UE}}(N)$.

Thus, higher type UEs receive more utility than the UEs whose types are lower. From the IC constraint and the two lemmas that we proved, we can easily deduce the following. If a high-type UE selects the contract designed for a low-type UE, even though a smaller transmission data rate is required from the BS, less reward received will deteriorate UE's utility. Moreover, if a lower type UE selects a contract intended for a high-type UE, the gain in terms of rewards cannot compensate the cost in power consumption for the high transmission data rate, and thus, the cost surpasses the gain. The UE can receive the maximum utility if and only if it selects the contract that best fit into its preference. Thus, we can guarantee that the contract is self reveal.

2.4.2 Optimal Contract

Given the contract feasibility constraints, we will formulate the system optimization problem in both discrete-type case and continuum-type case in this subsection.

2.4.2.1 Case of Discrete Type

Under the information asymmetry, the only information available at the BS is the probability λ_i with which a certain UE might belong to type θ_i. Our main focus is to maximize the utility of the BS, which represents the increased data rate when D2D communication is underlaid. Therefore, the problem can be posed as the following maximization

$$\max_{(T,R)} \sum_{i=1}^{N} \lambda_i \left(R_i - cT_i \right), \tag{2.18}$$

$s.t.$

$(a)\ \theta_i v(T_i) - R_i \geq 0,$

$(b)\ \theta_i v(T_i) - R_i \geq \theta_i v(T_j) - R_j,$

$(c)\ 0 \leq T_1 < \cdots < T_i < \cdots < T_N,$

$\quad i, j \in \{1, \cdots, N\}, \quad i \neq j.$

(a) and (b) represent the IR and IC constraints, respectively, and (c) represents the monotonicity condition. This problem is not a convex optimization problem; however, we can perform the following steps to find a solution:

Step 1: Reduce IR constraints. From (2.18), we can see that in total there are N IR constraints be satisfied. However, from Definition 2.1 we know that $\theta_1 < \cdots < \theta_i < \cdots < \theta_N$. By using IC constraints, we have

$$\theta_i v(T_i) - R_i \geq \theta_i v(T_1) - R_1 \geq \theta_1 v(T_1) - R_1 \geq 0. \tag{2.19}$$

Thus, if the IR constraint of *type-1* user is satisfied, the other IR constraints will automatically hold. Therefore, we only need to keep the first IR constraints and reduce the others.

Step 2: Reduce IC constraints. The IC constraints between *type-i* and *type-j*, $j \in \{1, \cdots, i-1\}$ are called downward incentive constraints (DICs). In particular, the IC constraint between *type-i* and *type-(i-1)* is called local downward incentive constraints (LDICs). Similarly, the IC constraints between *type-i* and *type-j*, $j \in \{i+1, \cdots, N\}$ are called upward incentive constraints (UICs), and the IC constraint between *type-i* and *type-(i+1)* is called local upward incentive constraints (LUICs). First, we prove that DICs can be reduced.

Proof As the number of users is N in our model, there exist $N(N-1)$ IC constraints in total. Here, we consider three types of users which follows $\theta_{i-1} < \theta_i < \theta_{i+1}$. Then, we have the following two LDICs

$$\theta_{i+1} v(T_{i+1}) - R_{i+1} \geq \theta_{i+1} v(T_i) - R_i \quad \text{and} \tag{2.20}$$
$$\theta_i v(T_i) - R_i \geq \theta_i v(T_{i-1}) - R_{i-1}. \tag{2.21}$$

In Lemma 2.1, we have shown that $T_i \geq T_j$ whenever $\theta_i \geq \theta_j > 0$, the second inequality becomes

$$\theta_{i+1}[v(T_i) - v(T_{i-1})] \geq \theta_i[v(T_i) - v(T_{i-1})] \geq R_i - R_{i-1} \quad \text{and} \tag{2.22}$$
$$\theta_{i+1} v(T_{i+1}) - R_{i+1} \geq \theta_{i+1} v(T_i) - R_i \geq \theta_{i+1} v(T_{i-1}) - R_{i-1}. \tag{2.23}$$

Thus, we have

$$\theta_{i+1} v(T_{i+1}) - R_{i+1} \geq \theta_{i+1} v(T_{i-1}) - R_{i-1}. \tag{2.24}$$

Therefore, if for *type-i* UE the LDIC holds, the incentive constraint with respect to *type-(i-1)* UE holds. This process can be extended downward from *type i − 1* to 1 UEs prove that all the DICs hold,

$$\theta_{i+1}v(T_{i+1}) - R_{i+1} \geq \theta_{i+1}v(T_{i-1}) - R_{i-1} \qquad (2.25)$$
$$\geq \cdots$$
$$\geq \theta_{i+1}v(T_1) - R_1,$$
$$N > i \geq 1.$$

Thus, we have completed the proof that with the LDIC constraint, all the DICs hold, that is,

$$\theta_i v(T_i) - R_i \geq \theta_i v(T_j) - R_j, \quad N \geq i > j \geq 1. \qquad (2.26)$$

Second, we prove all the UICs can be reduced.

Proof From the IC constraint, we have the following two LUICs:

$$\theta_{i-1}v(T_{i-1}) - R_{i-1} \geq \theta_{i-1}v(T_i) - R_i \quad \text{and} \qquad (2.27)$$
$$\theta_i v(T_i) - R_i \geq \theta_i v(T_{i+1}) - R_{i+1}. \qquad (2.28)$$

In Lemma 2.1 we have shown that $T_i \geq T_j$ whenever $\theta_i \geq \theta_j > 0$, the second inequality can be derived as

$$R_{i+1} - R_i \geq \theta_i(v(T_{i+1}) - v(T_i)) \geq \theta_{i-1}(v(T_{i+1}) - v(T_i)) \quad \text{and} \quad (2.29)$$
$$\theta_{i-1}v(T_{i-1}) - R_{i-1} \geq \theta_{i-1}v(T_i) - R_i \geq \theta_{i-1}v(T_{i+1}) - R_{i+1}. \qquad (2.30)$$

Thus, we have

$$\theta_{i-1}v(T_{i-1}) - R_{i-1} \geq \theta_{i-1}v(T_{i+1}) - R_{i+1}. \qquad (2.31)$$

Therefore, if for $type - (i - 1)$ UE, the incentive constraint with respect to $type - i$ UE holds, then all UICs are also satisfied. This process can be extended upward from $type\ i + 1$ to N UEs prove that all the UICs hold,

$$\theta_{i-1}v(T_{i-1}) - R_{i-1} \geq \theta_{i-1}v(T_{i+1}) - R_{i+1} \qquad (2.32)$$
$$\geq \cdots$$
$$\geq \theta_{i-1}v(T_N) - R_N,$$
$$N \geq i > 1.$$

Thus, we have complete the proof that with the LUIC constraint, all the UICs hold, that is

$$\theta_i v(T_i) - R_i \geq \theta_i v(T_j) - R_j, \quad 1 \leq i < j \leq N. \qquad (2.33)$$

Indeed, with the monotonicity condition $T_{i-1} < T_i$, the LDIC,

$$\theta_i v(T_i) - R_i \geq \theta_i v(T_{i-1}) - R_{i-1},$$ (2.34)

can easily imply that the LUIC,

$$\theta_{i-1} v(T_i) - R_i \leq \theta_{i-1} v(T_{i-1}) - R_{i-1},$$ (2.35)

can be satisfied and thus can be reduced. Thus, we have proved that, with the LDIC, all the UICs are reduced.

Step 3: Solve the optimization problem with reduced constraints. Thus, we can reduce the set of UICs and DICs, and only the set of LDICs and monotonicity condition are binding. Therefore, the optimization problem reduces to

$$\max_{(T,R)} \sum_{i=1}^{N} \lambda_i (R_i - cT_i),$$ (2.36)

$s.t.$

  (a) $\theta_1 v(T_1) - R_1 = 0,$

  (b) $\theta_i v(T_i) - R_i = \theta_i v(T_{i-1}) - R_{i-1},$

  (c) $0 \leq T_1 < \cdots < T_i < \cdots < T_N,$

    $i \in \{1, \cdots, N\}.$

To solve this problem, we can first formulate and solve the relaxed problem without the monotonicity condition and then consider the standard procedure of the Lagrangian multiplier. Then, we check whether the solution to this relaxed problem satisfies the monotonicity condition or not (Bolton and Dewatripont 2004).

The optimal contract solved by this optimization problem will give zeros utility for the lowest type of UEs. If $N = 2$, there are only two types of UEs, the high-type and the low type. By solving this optimization problem, the low-type UEs will obtain a zero utility contract and the high-type UEs can receive a positive utility. In general cases when $N > 2$, a similar conclusion is also provided in (Bolton and Dewatripont 2004; Gao et al. 2011, 2014), all types of UEs will get a positive utility except the lowest type UE who will get a zero utility.

2.4.2.2 Case of Continuum Type

In the previous case, there are N types of UEs from θ_1 to θ_N. In practice, the number of UEs types can be infinite. In this subsection, we will give an analysis about the continuum-type case with type θ which has the probability density function (PDF) $f(\theta)$ (with cumulative distribution function (CDF) $F(\theta)$ on the interval $[\underline{\theta}, \bar{\theta}]$. The contract that a BS offers to the UE is written as $[T(\theta), R(\theta)]$. T is monotonously increasing in R as in the discrete case. If no trading happens between the BS and the

UE, the contract is set as $T(\theta) = 0$ and $R(\theta) = 0$. Similar to the discrete-type case, we can write the BS's optimization problem as follows.

$$\max_{\{T(\theta), R(\theta)\}} \int_{\underline{\theta}}^{\overline{\theta}} [R(\theta) - cT(\theta)] f(\theta) d\theta, \tag{2.37}$$

$s.t.$

$$(a)\ \ \theta v[T(\theta)] - R(\theta) \geq 0,$$
$$(b)\ \ \theta v[T(\theta)] - R(\theta) \geq \theta v[T(\widehat{\theta})] - R(\widehat{\theta}),$$
$$\theta, \widehat{\theta} \in [\underline{\theta}, \overline{\theta}].$$

Condition (a) is the IR constraints and (b) represents the IC constraints. To solve this continuum-type case problem, we follow a similar process as the discrete-type case and begin by reducing the IR and IC constraints.

Step 1: Reduce IR constraints. We first reduce the number of IR constraints as did in the discrete case. Since the IC constraints hold, we have

$$\theta v[T(\theta)] - R(\theta) \geq \theta v[T(\underline{\theta})] - R(\underline{\theta}) \tag{2.38}$$
$$\geq \underline{\theta} v[T(\underline{\theta})] - R(\underline{\theta}).$$

Thus, if the IR constraint of $\underline{\theta}$ is satisfied, the IR constraints for all the other values of θ will automatically hold. Therefore, replace the IR constraints by

$$\underline{\theta} v[T(\underline{\theta})] - R(\underline{\theta}) \geq 0. \tag{2.39}$$

Step 2: Reduce IC constraints. To reduce the IC constraints, we give Lemma 2.3 that using two other constraints to replace all IC constraints (Bolton and Dewatripont 2004).

Lemma 2.3 *The IC constraint is equivalent to the following two conditions:*

1. *Monotonicity*

$$\frac{dT(\theta)}{d\theta} \geq 0. \tag{2.40}$$

2. *Local incentive compatibility*

$$\theta v'[T(\theta)] \frac{dT(\theta)}{d\theta} = R'(\theta), \theta \in [\underline{\theta}, \overline{\theta}]. \tag{2.41}$$

Proof The monotonicity can be easily derived following the steps in Lemma 2.1 and Definition 2.4. The local incentive compatibility can be proved by contradiction. Suppose we have the monotonicity and local incentive compatibility, and the IC constraint cannot be held. Then, with at least one $\widehat{\theta}$ violates the IC constraint

$$0 \leq \theta v[T(\theta)] - R(\theta) < \theta v[T(\widehat{\theta})] - R(\widehat{\theta}). \tag{2.42}$$

Integrating it from θ to $\widehat{\theta}$, we get

$$\int_{\theta}^{\widehat{\theta}} \left[\theta v'[T(x)]\frac{dT(x)}{dx} - R'(x) \right] dx > 0. \tag{2.43}$$

From the local incentive compatibility, we know $\int_{\theta}^{\widehat{\theta}} \left[xv'[T(x)]\frac{dT(x)}{dx} - R'(x) \right] dx = 0$. If $\theta < x < \widehat{\theta}$, from the monotonicity we have $\theta\frac{dv(T(x))}{dx} \le x\frac{dv(T(x))}{dx}$. Therefore,

$$\int_{\theta}^{\widehat{\theta}} \left[\theta v'[T(x)]\frac{dT(x)}{dx} - R'(x) \right] dx < 0. \tag{2.44}$$

Thus, we see a contradiction. Similarly, if $\theta > \widehat{\theta}$, we can also get a contradiction. Thus, the two conditions, monotonicity and local incentive compatibility, can guarantee the UE's incentive compatible constraints.

Step 3: Optimization problem with reduced constraints. Finally, the BS's optimization problem can be written as

$$\max_{(T(\theta), R(\theta))} \int_{\underline{\theta}}^{\overline{\theta}} [R(\theta) - cT(\theta)] f(\theta)d\theta, \tag{2.45}$$

$s.t.$

$$(a) \quad \underline{\theta}v[T(\underline{\theta})] - R(\underline{\theta}) \ge 0,$$

$$(b) \quad \theta v'[T(\theta)]\frac{dT(\theta)}{d\theta} = R'(\theta),$$

$$(c) \quad \frac{dT(\theta)}{d\theta} \ge 0,$$

$$\theta \in [\underline{\theta}, \overline{\theta}].$$

Similar to the discrete-type case problem, constraints (a) and (b) represent the IR and IC constraints, and constraint (c) is the monotonicity condition. The procedure for solving this problem is also similar to the discrete-type case problem. First ignore the monotonicity condition and solve the relaxed problem with constraints (a) and (b). Then, check whether the solution to this relaxed problem satisfies the monotonicity condition or not.

2.4.3 Practical Implementation

By solving the proposed problem, we could provide UEs with the optimal contract that can incentivize them to participate in D2D communication. To implement the proposed approach in a practical D2D network, we can follow the next steps. From the system model, we have the initial information such as the cellular network radius S, the cellular users' transmit power P_c, the number of UE types N, and the probability λ_i that UE belongs to θ_i. With those initial values, the BS can obtain the optimal contract (T, R). Once there are UEs requesting contents, the BS acts in the following stages.

In the first stage, when the BS receives UEs' requests for contents, the BS will detect if the contents are locally accessible in other UEs within the maximum D2D communication distance L. If the content is locally available, then the BS will broadcast the optimal contracts to the candidate content providers. By evaluating the contracts, UEs will send feedback signals to indicate whether they are willing to participate in according to the estimated utility. After getting feedback from UEs, the BS will sign the contract with the UE that accepts it. If all UEs reject the contract, the BS will serve the content requester directly, which is the same procedure as if the content is not locally accessible.

After signing the contract, the employed UE will set up the D2D communication and forward the content to the content requester. The BS will stand by to watch the communication by sending control signals and also receiving feedback signals from UEs. If the transmission is successful, the BS rewards the involved UEs based on their contract. Otherwise, if the transmission failed, the BS serves the user directly and the "employed" UE will not receive the reward. The proposed D2D communication algorithm is summarized in Algorithm 1. This algorithm gives the practical implementation steps of the theoretical model.

2.5   Simulation Results and Analysis

In this part, we will first evaluate the feasibility of the proposed contract and then analyze the system performance when D2D communication is underlaid in the cellular network.

First of all, we donate the optimal contract solved in the previous section by *information asymmetry*. For comparison purposes, we introduce another two incentive mechanisms. The first one is the optimal contract under *no information asymmetry* (i.e., the BS is aware of the types of UEs), which is the optimal outcome that we can achieve and serve as the upper bound. The second contract is the *linear pricing* which is also under the information asymmetry that the BS has no acknowledgment of the UE type. In this *linear pricing* mechanism, the BS will only specify a unit price P for data rate, and the UEs will request the amount of reward T which corresponding to a certain amount of data rate, to maximize their own utilities.

We assume $N = 20$ and give the simulation with 20 types of UEs. For simplicity, we consider a uniform distribution of UE type, i.e., $\lambda_i = 1/N$. We set the unit payment cost of the BS $c = 0.01$. The main parameters of the D2D underlaid cellular network are shown in Table 2.1.

Table 2.1 Physical layer parameters

Parameter	Value
Cellular area radius	500 m
Maximum D2D distance	30 m
number of UE types	20
Noise spectral density	−174 dBm/Hz
Noise figure	9 dB at device
Antenna gains	BS: 14 dBi; device: 0 dBi
Transmit power	BS: 46 dBm; device: 23 dBm

2.5.1   Contract Feasibility

2.5.1.1   Monotonicity

In Fig. 2.2a, b, we compare the required transmission data rate and reward of different type UEs to show the monotonicity of the contract.

In Fig. 2.2a, we see that required transmission data rate increases with the UE type, which is consistent with our system model. The difference among the three mechanisms is that the required data rate under *no information asymmetry* and *linear pricing* is linear function of type and is a concave function of type under *information asymmetry*. Among the three mechanisms, the *no information asymmetry* contract requires the highest data rate from the UE, followed by the optimal contract under *information asymmetry*. The lowest data rate is required under the *linear pricing* contract. Similarly, the reward shown in Fig. 2.2b also proves our assumption that reward T is a strictly increasing function of UE type.

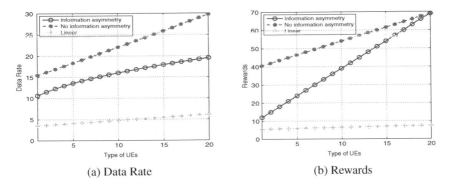

(a) Data Rate   (b) Rewards

Fig. 2.2   Contract monotonicity

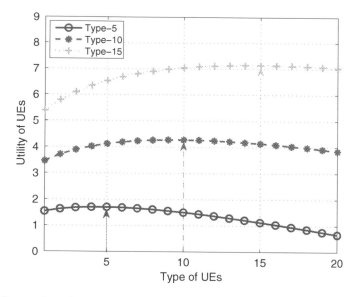

Fig. 2.3 Contract incentive compatibility

2.5.1.2 Incentive Compatibility

In Fig. 2.3, we evaluate the incentive compatibility of our proposed contract, the optimal scheme. We show the utilities of *type-5*, *type-10*, and *type-15* UEs when selecting all the contracts offered by the BS. The utility of each user is a concave function. Each UE can achieve their maximum utility if and only if it selects the type of contract that is intended for its own type, as shown clearly in Fig. 2.3. Thus, by designing a contract in this form, the type of an UE will be automatically revealed to the BS after its selection. In other words, the optimal contract under *information asymmetry* enables that the BS breaks the information asymmetry and retrieves the information related to UE type.

Moreover, Fig. 2.3 shows that when the three types of users select the same contracts, their utilities follow the inequality $u_5 < u_{10} < u_{15}$. This corroborates the result shown by the (2.16) in Lemma 2.2: the higher the type of the UE, the larger the utility it can receive when selecting the same contract.

2.5.2 System Performance

To evaluate the performance of the D2D underlaid cellular network, we try to see the impacts of different parameters on the utility of BS, UE, and social welfare.

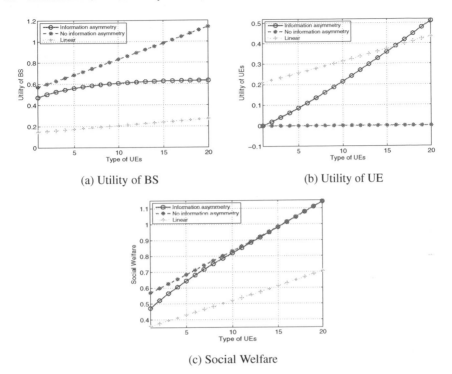

(a) Utility of BS   (b) Utility of UE

(c) Social Welfare

Fig. 2.4 System performance of different types of UEs

2.5.2.1 The UE Type

First, we take a close look at the three values of different types of UEs in Fig. 2.4. The three figures show the monotonicity of the contract that the higher the UE type, the larger the utility it will bring to the BS and UE, as well as the social welfare.

Figure 2.4a shows that the BS achieves the highest utility when there is no information asymmetry, since the BS has full knowledge of UE types. Nonetheless, we can see that the proposed solution with information asymmetry yields a utility for the BS that outperforms the linear pricing case. Here, we note that even though the optimal contract under *information asymmetry* can force the UEs to reveal their types, the exact value of the UE type is still unavailable to the BS. Thus, the BS can only achieve a near-optimal utility under *information asymmetry*, which is always upper bounded by the *no information asymmetry* case. The *linear pricing* mechanism does not place any restriction on the UEs choice of contract and less information is retrieved, which prevents the BS from obtaining more utility.

In Fig. 2.4b, we compare 20 types of UEs' utilities. These results proved the monotonicity of the contract that the higher the type of UE, the larger the utility it can receive under *information asymmetry*. All the types of UEs enjoy a positive utility except the lowest type (i.e., *type-1*) UE, which is consistent with our conclusion in

Sect. 2.4.2.2. However, the UE's utility remains 0 disregarding the type of UE under *no information asymmetry*. This is due to the fact that when the BS is available at the UE's type, it will adjust the contract to maximize its own utility while leave the UE a 0 utility. Overall, we see that *linear pricing* gives the UEs the highest utility, followed by the optimal contract under *information asymmetry*, then the ideal case with *no information asymmetry*. However, for some of the high-type UEs can obtain higher utility from the optimal contract under *information asymmetry* than the *linear pricing*.

In Fig. 2.4c, we see that the social welfare shows similar performance with that of the BS. One interesting point is that, the social welfare of the highest type UE has the same value under *no information asymmetry* and *information asymmetry*. This is in accordance with the conclusion we made in Sect. 2.4.2.2 that the highest type UE will result in an efficient trading as if there is no information asymmetry. For other high-type UEs under *no information asymmetry*, they also have close optimal efficient trading with the BS. The *linear pricing* mechanism gives the lowest social welfare (i.e., trading efficiency) since no information retrieving strategy has been applied.

2.5.2.2 The Cellular Network Size

In a small-sized network, cellular communication will generate severe interference on D2D communication, which will decrease the transmission data rate of UEs. The interference will decrease as the size of network increases. In Fig. 2.5, we show the impact of network size on the system's performance.

In Fig. 2.5a, b, we show the utility of the BS and UEs when the cellular network size varies, when the transmission power and the antenna gain of the BS are fixed. As the size of cellular decreases, D2D UE pairs and cellular UEs are located in a more dense area, and suffering from a larger interference from other cellular and D2D UEs. Thus, the transmission data rate decreases, as well as the rewards. As a result, the utilities of the BS and UE also decrease.

From Fig. 2.5a, we see that the utility of BS achieves the maximum utility under *no information asymmetry*, followed by the optimal contract under *information asymmetry*. The *linear pricing* gives the worst utility to the BS which compares to the other two. The utility of the UE has one similar property as Fig. 2.4b that the UE utility under *no information asymmetry* remains 0. The UE achieves the maximum utility by the *linear pricing*, followed by the optimal contract under *information asymmetry*. The UEs benefit from the information asymmetry, while the BS can increase its utility by removing the information asymmetry.

From Fig. 2.5c, we can also see the differences in the social welfare under the three different contracts. Social welfare under *no information asymmetry* achieves the highest among the other two. As the BS is informed of the UE type, the transaction achieves the highest efficiency, then followed by the optimal contract achieved under *information asymmetry*. The *linear pricing* presents the worst efficiency. The optimal contract achieved under *information asymmetry* achieves a near-optimal social

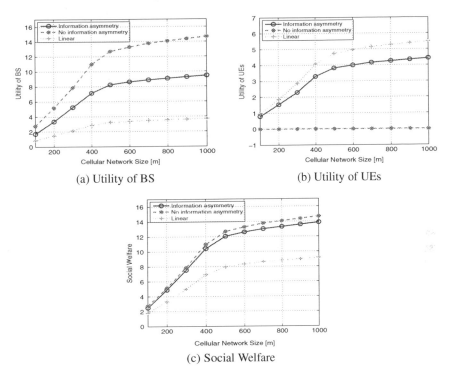

Fig. 2.5 The system performance when the size of cellular network varies

welfare, as it breaks the information asymmetry when the UEs select contracts, their types are revealed to the BS automatically. The *linear pricing* does not account for any type of information and thus has the lowest social welfare.

2.5.2.3 The Maximum D2D Communication Distance

When the size of the cellular network and the BS transmission power are fixed, the interference from cellular communication will be in a certain range. Under this condition, we change the maximum transmission distance of D2D pairs, to see the effects on system performance, in Fig. 2.6.

For the utility of the BS and UEs, Fig. 2.6a, b still exhibit similar properties as shown in Fig. 2.5a, b. The utility that the BS receives is maximized under *no information asymmetry*, followed by *information asymmetry* and *linear pricing*. The UE achieves the maximum utility under *linear pricing*, followed by *information asymmetry* and *no information asymmetry* which equals to 0 all the time. The highest social welfare is achieved under *no information asymmetry*, *information asymmetry* is the second, and *linear pricing* results in the worst social welfare.

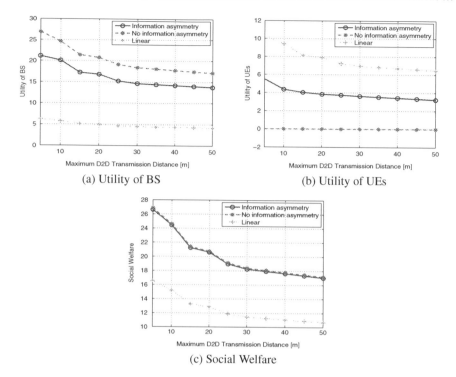

Fig. 2.6 The system performance when the maximum D2D communication distance varies

2.5.2.4   The Number of UE Types

In Fig. 2.7, we study the system performance when the number of UE types increases, while the other parameters are fixed. An increase in the number of types will automatically yield an increase in the total number of UEs pairs. Thus, the utilities of the BS and UE and the social welfare will also increase.

Similar to the conclusions drawn from the Figs. 2.5 and 2.6, the BS has the highest utility under *no information asymmetry*. The optimal contract under *information asymmetry* gives the second highest BS utility. The *linear pricing* still gives the worst utility to the BS. The *linear pricing* gives the highest UE utility, the optimal contract under *information asymmetry* gives the second highest one, and the *no information asymmetry* remains 0. The case under *no information asymmetry* achieves the highest social welfare among all schemes. The optimal contract under *information asymmetry* yields the second highest social welfare. The *linear pricing* still achieves the lowest efficiency in social welfare.

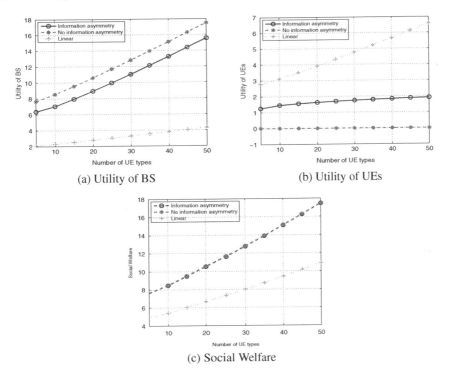

(a) Utility of BS

(b) Utility of UEs

(c) Social Welfare

Fig. 2.7 The system performance when the number of UE types varies

2.6 Summary

In this chapter, we have proposed an *adverse selection* model for addressing the
problem of incentivizing UEs to participate in D2D communication underlaid over
a cellular system. Under the case with information asymmetry in which the UEs'
preferences are not available at the BS, we have proposed a self-revealing mechanism
based on the framework of contract theory. We have considered the type of UEs under
two different scenarios, the discrete-type case and continuum-type case. Simulation
results have shown that our proposed approach can potentially incentivize UEs to
participate in D2D communication. Furthermore, the optimal contract under *information asymmetry* has been proved to obtain the performance close to the ideal case
with *no information asymmetry*, and higher than the *linear pricing* when not trying
to retrieve any information at all. As there are many literatures have also covered
the *adverse selection* problem in wireless networks, we will start to talk about other
contract-theoretical models, such as *moral hazard*, mixed problem, and incomplete
contract, which are less covered in the previous literature from the next chapter.

References

Bolton P, Dewatripont M (2004) Contract theory. The MIT Press, Cambridge, MA

Camps-Mur D, Garcia-Saavedra A, Serrano P (2013) Device-to-device communications with Wi-Fi direct: overview and experimentation. IEEE Wirel Commun 20(3):96–104

Cisco (2011) Cisco visual networking index: global mobile data traffic forecast update. Whitepaper 10(8):2752–2763

Duan L, Gao L, Huang J (2014a) Cooperative spectrum sharing: a contract-based approach. IEEE Trans. Mobile Comput. 13(1):174–187

Duan L, Kubo T, Sugiyama K, Huang J, Hasegawa T, Walrand J (2014b) Motivating smartphone collaboration in data acquisition and distributed computing. IEEE Trans. Mobile Comput. 13(10):2320–2333

Feng D, Lu L, Yuan-Wu Y, Li GY, Feng G, Li S (2013) Device-to-device communications underlaying cellular networks. IEEE Trans. Commun. 61(8):3541–3551

Gao L, Wang X, Xu Y, Zhang Q (2011) Spectrum trading in cognitive radio networks: a contract-theoretic modeling approach. IEEE J Sel Areas Commun (JSAC) 29(4):843–855

Gao L, Huang J, Chen Y, Shou B (2013) An integrated contract and auction design for secondary spectrum trading. IEEE J Sel Areas Commun (JSAC) 31(3):581–592

Gao L, Huang J, Chen Y, Shou B (2014) Contract-based cooperative spectrum sharing. IEEE Trans Mobile Comput 13(1):174–187

Golrezaei N, Molisch AF, Dimakis AG, Caire G (2013) Femtocaching and device-to-device collaboration: a new architecture for wireless video distribution. IEEE Commun Mag 51(4):142–149

Jin H, Sun G, Wang X, Zhang Q (2012) Spectrum trading with insurance in cognitive radio networks. In: The 31st IEEE international conference on computer communications (INFOCOM), Orlando, FL

Knapper R, Blau B, Conte T, Sailer A, Kochut A, Mohindra A (2011) Efficient contracting in cloud service markets with asymmetric information—a screening approach. In: 2011 IEEE 13th conference on commerce and enterprise computing (CEC), Luxembourg

Min H, Lee J, Park S, Hong D (2011a) Capacity enhancement using an interference limited area for device-to-device uplink underlaying cellular networks. IEEE Trans Wirel Commun 10(12):3995–4000

Min H, Lee J, Park S, Hong D (2011b) Capacity enhancement using an interference limited area for device-to-device uplink underlaying cellular networks. IEEE Trans Wirel Commun 10(12):3995–4000

Min H, Seo W, Lee J, Park S, Hong D (2011c) Reliability improvement using receive mode selection in the device-to-device uplink period underlaying cellular networks. IEEE Trans Wirel Commun 10(2):413–418

Qualcomm (2012) LTE direct: research and use cases. Technical report. http://www.qualcomm.com/about/research/projects/lte-direct

Quek TQS, de la Roche G, Guvenc I, Kountouris M (2013) Small Cell Networks: Deployment, PHY Techniques, and Resource Management. Cambridge University Press, UK

Sesia S, Toufik I, Baker M (2009) LTE: the UMTS long term evolution. John Wiley & Sons, New York

Song L, Niyato D, Han Z, Hossain E (2014) Wireless device-to-device communications and networks. Cambridge University Press, UK

Tanbourgi R, Jakel H, Jondral FK (2014) Cooperative interference cancellation using device-to-device communications. IEEE Commun Mag 52(6):118–124

Werin L, Wijkander H (1992) Contract economics. Blackwell Publishers, Oxford, UK

Xing B, Seada K, Venkatasubramanian N (2009) An experimental study on Wi-Fi ad-hoc mode for mobile device-to-device video delivery. In: The 28th IEEE international conference on computer communications (INFOCOM), Rio de Janeiro, Brazil

Xu C, Song L, Han Z, Li D, Jiao B (2012) Resource allocation using a reverse iterative combina-
    torial auction for device-to-device underlay cellular networks. In: IEEE globe communication
    conference (GLOBECOM), Anaheim, CA
Xu C, Song L, Han Z (2014) Resource management for device-to-device underlay communication.
    Springer, Germany
Yaacoub E (2014) On the use of device-to-device communications for qos and data rate enhancement
    in LTE public safety networks. In: IEEE WCNC—Workshop on device-to-device and public
    safety communications, Istanbul, Turkey
Yu CH, Doppler K, Ribeiro CB, Tirkkonen O (2011) Resource sharing optimization for device-to-
    device communication underlaying cellular networks. IEEE Trans Wirel Commun 10(8):2752–
    2763
Zhang R, Cheng X, Yang L, Jiao B (2013) Interference-aware graph based resource sharing for
    device-to-device communications underlaying cellular networks. In: Wireless communications
    and networking conference (WCNC). IEEE, Shanghai, China
Zulhasnine M, Huang C, Srinivasan A (2010) Efficient resource allocation for device to device
    communication underlaying LTE network. In: The 6th IEEE international conference on wireless
    and mobile computing networking and communications (WiMob), Niagara Falls, Canada

Chapter 3
Incentive Mechanism in Crowdsourcing with Moral Hazard

3.1 Introduction

From this chapter, we will discuss several applications of the *moral hazard* problem by adopting different models. One fundamental model will be given in this chapter as an introductory. Then, two extensions will be discussed sequentially. Nowadays, people are used to access various sophisticated location-based services (e.g., Foursquare and Yelp) from their smartphones through wireless access networks. Most location-based services require user to regularly transmit data to the principal, which is obtained by the embedded sensor in smartphone. One brief illustration of crowdsourcing is shown in Fig. 3.1, in which the smartphone users regularly transmit data to the principal. Once the data is processed by the principal, the location-based service is provided to the users for free. One well-known application is the live auto traffic map offered by Google. Smartphone users transmit traffic information which includes the time, location, and velocity to Google. Google collects those data and processes necessary data analysis before providing the free live traffic map to the users (Angmin and Valentino-Devries 2011).

With the drastic growth in the global location-based service market, as well as the rapid development of big data technology, more data as well as user participation are required to support more sophisticated services. While participating in a crowdsourcing activity, smartphone users consume their own resources such as battery and computing power. In addition, users expose their locations with potential privacy threats (Zhao et al. 2014). Many users are hesitated to participate in with those concerns, which become one of the serious impediments to the development of location-based services. Thus, necessary incentive mechanisms that motivate the users to participate are needed to address the new demands.

Many researches have already noticed that there is an urgent need to alleviate the conflict by introducing an incentive mechanism to the users. The plan proposed by Karma (2012) is to reward users with a fixed reward for the first time of participating in, while the problem of this mechanism is lacking of continuous incentive for users to stay actively after receiving the opening reward (Gao et al. 2014). Inspired by

© Springer International Publishing AG 2017
Y. Zhang and Z. Han, *Contract Theory for Wireless Networks*,
Wireless Networks, DOI 10.1007/978-3-319-53288-2_3

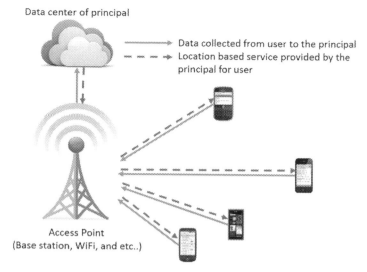

Data center of principal

Data collected from user to the principal
Location based service provided by the principal for user

Access Point
(Base station, WiFi, and etc..)

Fig. 3.1 An illustration of crowdsourcing

the effort-based reward from the labor market, several works have been proposed to address this problem by providing a reward-performance consistent contract. The work in Luo et al. (2014) derives the optimal contribution-dependent price function that induces the maximum profit for the principal, while strictly maintaining the incentive for users to participate in the Bayesian game equilibrium. Similarly, based on contract theory, Duan et al. (2012) offers an effort-based contract for users to select the amount of work they would like to accomplish voluntarily.

In this chapter, we adopt the performance-related contract to incentivize users to turn on their sensors and allow data collecting for the principal. We assume that the principal will offer a compensation package based on the performance that user achieves. The performance can be measured by the quantity, quality, or importance of the data. From the principal's aspect, it makes profit by extracting useful information from the collected data and by selling advertisements embedded in the location-based service, while the principal also incurs a cost such as the rewards for users and the operation cost. Thus, the principal needs to find a proper mechanism that can maximize its own benefit, while at the same time ensures the users receive necessary incentives for participation.

The moral hazard model of contract theory provides us a useful tool to design such a mechanism, which aims at solving the employees' incentive problems when performing a task. Moral hazard arises when employees are not taking the full responsibilities for their tasks, and they are intending to shirk and act less carefully during work (Werin and Wijkander 1992). In our model, the principal is facing the moral hazard of the users. Users like to enjoy the location-based service provided by the principal, but dislike exposing their location and collecting location-based information for the principal.

Indeed, the moral hazard model can be adopted fluently in the crowdsourcing activity. First, the principal can specify a performance-related compensation package to the user. Based on the compensation scheme, the user will then maximize its own utility by deciding how much effort it will contribute to the crowdsourcing process. The key point is to find out the compensation package that satisfies both the principal and the user. In a nutshell, we need to design a compensation package in which the users will select the amount of effort in accordance with the compensation received from the principal, and the principal's utility is maximized.

The main contributions of this chapter are as follows: First, we adopt a performance-related compensation package to provide users with a continuous incentive to participate in crowdsourcing activity. Second, we propose a novel approach to solve the optimal compensation package using the framework of moral hazard, which is rooted in economic research. Last, through simulations, we provide a thorough study of the key parameters' influences on the compensation package. In addition, we propose three other mechanisms to serve as the comparisons, which prove the effectiveness of the proposed mechanism and show that the users obtain continuous incentives to participate in crowdsourcing.

The remainder of this chapter is organized as follows: First, we will introduce the network model in Sect. 3.2. Then, the problem formulation is described in Sect. 3.3, and we propose the solution of the optimization problem. The performance evaluation is conducted in Sect. 3.4. Finally, Sect. 3.5 gives the summary of this chapter.

3.2 System Model

We consider a user who participates in a crowdsourcing activity initiated by a principal. In this section, we will first construct the compensation package offered by the principal, and then give the utility functions of both the user and principal before proceeding to the solution of the optimal compensation package.

3.2.1 Utility of User

Inspired by the manager's compensation package in industry, which comprises a fixed salary, a bonus related to the firm's profits in the current year, and stock options related compensation based on the firm's share price (Bebchuk et al. 2002), we define the compensation package w in a crowdsourcing activity as a combination of a fixed salary t, a short-term bonus, and a long-term bonus. The short-term bonus is the reward related to the user's current performance which can be referred to the quality of the received data q (e.g., quantity, quality, correlativeness, and importance of the data). The long-term bonus can be regarded as the user's benefit from using the location-based service provided by the principal for free, and this benefit is highly correlated with the quality of the location-based service p (e.g., update frequency, effectiveness, reliability, popularity, privacy, and security of the service). Thus, we can also regard the quality of location-based service as the performance of the principal.

When user helps to collect useful data for the principal, we regarded it as the user is making an effort $a \in A$, while the effort a is hidden from the principal who can only observe the performance q, i.e., the quality of the received data. Due to some measurement errors, the quality of received data q is slightly different from the actual effort exerted by the user. Therefore, the performance of the user is a noisy signal of its effort. Thus, we assume that the quality of received data q to be normally distributed with mean a and variance σ_q^2:

$$q = a + \varepsilon_q, \tag{3.1}$$

where $\varepsilon_q \sim N(0, \sigma_q^2)$.

Similarly, as the effort the user takes in the crowdsourcing is one significant feature that determines the quality of location-based service, we take the quality of location-based service p to be normally distributed with mean a and variance σ_p^2:

$$p = a + \varepsilon_p, \tag{3.2}$$

where ε_p is a normally distributed random variable $\varepsilon_p \sim N(0, \sigma_p^2)$. As the user's evaluation of the quality of the location-based service varies from person to person, a positive or negative ε_p can be regarded as the user's personal favor toward the location-based service.

Without loss of generosity, we assume that both the short- and long-term bonuses are linear contracts with the qualities of the received data and location-based service. By restricting the compensation package offered by the principal in the linear form, the compensation package w the user receives by participating in the crowdsourcing activity can be written as

$$w = t + sq + fp, \tag{3.3}$$

where t denotes the fixed compensation salary (independent of performance), s denotes the fraction of reward related to the user's performance q, and f is the fraction of reward related to the principal's performance p. Therefore, sq represents the short-term bonus (performance-related component of compensation) and fp means the long-term bonus from experiencing the free location-based service.

As the popularity of a location-based service depends on many aspects such as the market and app design, the qualities of receiving data q and location-based service p behave differently. While they are likely to be correlated random variables, the quality of the received data provides one of the solid supports for the success of a location-based service. Therefore, we let σ_{qp} denote the covariance of q and p. In practice, we rarely meet the situation where a high quality of the receiving data q corresponds to a low quality of location-based service p. In other words, the two variables q and p cannot be negatively correlated. Thus, we assume that the covariance σ_{qp} is a nonnegative value, with $\sigma_{qp} \geq 0$.

In this model, we assume that the user has constant absolute risk averse (CARA) risk preferences, which means the user has a constant attitude toward risk as its income increases. Thus, user utility is represented by a negative exponential utility form:

$$u(w, a) = -e^{-\eta[w-\psi(a)]},\tag{3.4}$$

where $\eta > 0$ is the user's coefficient of absolute risk aversion ($\eta = -u''/u'$). A larger value of $\eta > 0$ means less incentive for the user to implement an effort. $\psi(a)$ is the incurring cost in providing the effort a for the principal. For simplicity, the cost function is assumed to be quadratic:

$$\psi(a) = \frac{1}{2}ca^2.\tag{3.5}$$

The utility and cost of the user are measured in such monetary units that they are consistent with the compensation from the principal.

3.2.2 Utility of Principal

In this model, we regard the principal as a "buy and hold" investor, who cares only about the direct performance of the user (Tirole 1993). That is, the principal is not concerned about its profit from the location-based service in the secondary market (e.g., advertisement selling). Therefore, the utility of the principal is not directly concerned about the quality of the location-based service p. Thus, we define the utility of the principal as the evaluation of the quality of received data q minus the compensation package w to the user. Thus, the principal's utility is written as

$$V(w, a) = E(q - w),\tag{3.6}$$

where $E(\cdot)$ is the evaluation function that follows $E(0) = 0$, $E'(\cdot) > 0$, and $E''(\cdot) \geq 0$.

Different from the user who has CARA risk preferences, the principal here is assumed to be risk neutral, i.e., $E''(\cdot) = 0$. Thus, the utility of the principal can be simplified as

$$V(w, a) = q - w = (1 - s - f)a - t.\tag{3.7}$$

3.3 Problem Formulation

With utility definitions of the user and principal, we can solve the principal's utility maximization problem while providing the user with necessary incentives. The principal's problem can be written as

$$\max_{a,t,s,f} V(w, a), \tag{3.8}$$

$$s.t. \quad (a) \ \ a^* \in \arg\max_a u(w, a),$$

$$(b) \ \ u(w, a) \geq u(\overline{w}),$$

where $u(\overline{w})$ is the reservation utility of the user when not taking any effort ($a = 0$) in the crowdsourcing. The principal maximizes its own utility under the incentive compatible (IC) constraint (a) that the user selects the optimal effort that maximizes its own utility, and the individual rationality (IR) constraint (b) that user receives no less than its reservation utility.

Maximizing the user's expected utility $u(w, a)$ is equivalent to maximizing

$$u(w, a) = -e^{-\eta[t+s(a+\varepsilon_q)+f(a+\varepsilon_p)-\frac{1}{2}ca^2]}, \tag{3.9}$$

$$= -e^{-\eta(t+sa+fa-\frac{1}{2}ca^2)}e^{-\eta(s\varepsilon_q+f\varepsilon_p)}.$$

In (3.9), the first part of the utility function is a constant value, while the second part which includes random variables ε_q and ε_p needs further simplification. From (Bolton and Dewatripont 2004), we see that

$$e^{\gamma\varepsilon} = e^{\frac{\gamma^2\sigma^2}{2}}, \tag{3.10}$$

where γ is a constant and ε is a normally distributed random variable with $\varepsilon \sim N(0, \sigma^2)$. As we have mentioned previously, there are $\varepsilon_q \sim N(0, \sigma_q^2)$ and $\varepsilon_p \sim N(0, \sigma_p^2)$. We can assume a new random variable $\varepsilon' = s\varepsilon_q + f\varepsilon_p$. The new variable ε' still follows the normal distribution $N(0, \sigma'^2)$. The variance σ'^2 of ε' can be derived as follows:

$$Var(\varepsilon') = Var(s\varepsilon_q + f\varepsilon_p), \tag{3.11}$$

$$= Var(s\varepsilon_q) + Var(f\varepsilon_p) + 2Cov(s\varepsilon_q, f\varepsilon_p),$$

$$= s^2\sigma_q^2 + f^2\sigma_p^2 + 2sf\sigma_{qp}.$$

Thus, implementing (3.10) and (3.11) into (3.9), we have

$$u(w, a) = -e^{-\eta(t+sa+fa-\frac{1}{2}ca^2)+\frac{1}{2}\eta(s^2\sigma_q^2+2sf\sigma_{qp}+f^2\sigma_p^2)}. \tag{3.12}$$

With the definition of the principal's utility function in (3.7), we can rewrite the principal and user's objectives in terms of their certainty equivalent wealth and thus obtain the following simple representation of the principal's problem:

$$\max_{a,t,s,f} (1 - s - f)a - t, \tag{3.13}$$

s.t. (a) $a \in \arg\max_a[(s + f)a + t - \frac{1}{2}ca^2 - \frac{1}{2}\eta(s^2\sigma_q^2 + 2sf\sigma_{qp} + f^2\sigma_p^2)],$

(b) $[(s + f)a + t - \frac{1}{2}ca^2 - \frac{1}{2}\eta(s^2\sigma_q^2 + 2sf\sigma_{qp} + f^2\sigma_p^2)] \geq \overline{w},$

where $\overline{w}$ denotes the reservation compensation of the user when not participating in the crowdsourcing activity.

This problem can be solved by using the first-order approach. In the first step, we take the first derivative of the user's utility function regarding a and set it $u'(w, a) = 0$. Then, we obtain the optimal effort a^*

$$a^* = \frac{s + f}{c}. \tag{3.14}$$

Accordingly, we substitute the IR constraint with the optimal action a^* and simplify the owner's problem to

$$\max_{t,s,f}(1 - s - f)\frac{s + f}{c} - t, \tag{3.15}$$

s.t. (a) $[(s + f)\frac{s + f}{c} + t - \frac{1}{2}c\left(\frac{s + f}{c}\right)^2 - \frac{1}{2}\eta(s^2\sigma_q^2 + 2sf\sigma_{qp} + f^2\sigma_p^2)] = \overline{w}.$

Substituting for the value of t in the IR constraint and maximizing with respect to s and f, we then have the fraction of reward s^* related to performance and the fraction of benefit f^* from the location-based service in the optimal linear compensation package as follows:

$$s^* = \frac{\sigma_p^2 - \sigma_{qp}}{\sigma_q^2 + 2\sigma_{qp} + \sigma_p^2}\frac{1}{1 + \eta c\Omega}, \tag{3.16}$$

$$f^* = \frac{\sigma_q^2 - \sigma_{qp}}{\sigma_q^2 + 2\sigma_{qp} + \sigma_p^2}\frac{1}{1 + \eta c\Omega}, \tag{3.17}$$

where

$$\Omega = \frac{\sigma_q^2\sigma_p^2 - \sigma_{qp}^2}{\sigma_q^2 + 2\sigma_{qp} + \sigma_p^2}. \tag{3.18}$$

Representing t by $\overline{w}$, s, and f, we obtain the fixed salary t in the optimal linear compensation package as follows:

$$t^* = \overline{w} + \frac{1}{2}\eta(s^2\sigma_q^2 + 2sf\sigma_{qp} + f^2\sigma_p^2) + \frac{1}{2}c\left(\frac{s + f}{c}\right)^2 - (s + f)\frac{s + f}{c}. \tag{3.19}$$

These expressions determine how the user's compensation package varies as a function of the underlying environment in which the crowdsourcing activity goes on. Specifically, the compensation package can be directly tied to the stochastic structure of the qualities of received data and location-based service which are related to the user's effort.

3.4   Simulation Results and Analysis

In this section, we will first give an analysis about the compensation package obtained in the previous section by varying the parameters such as the cost coefficient, standard deviation, covariance, and risk averse degree. Then, we will conduct a comparison among several compensation packages in different scenarios. In the simulation setup, we assume that the reservation salary of the user $\overline{w} = 0$ when not participating in the crowdsourcing ($a = 0$). We notice that, in the optimal compensation package we have derived, no matter how those parameters change, the user's utility will remain the same. As the principal tries to maximize its own utility, it restricts the user's utility as small as possible, as long as the user can be incentivized to participate in the crowdsourcing. Thus, the optimal compensation package will bring user the utility the same as the reservation utility $-e^{-\eta \overline{w}}$, which in our case is -1 as we set $\overline{w} = 0$.

3.4.1   Optimal Compensation Package Analysis

In Fig. 3.2, we investigate the impacts of covariance σ_{qp} on the compensation package, while fixing the standard deviations σ_q and σ_p, cost coefficient c, and risk averse degree η. The simulation results show that as the covariance σ_{qp} increases, the optimal effort a, compensation package w, and the principal's utility are all decreasing. Meanwhile, we see that the fixed salary t, short-term bonus sq, and long-term bonus fp in the compensation package are decreasing as well. This result is because as the relationship between the qualities of received data and the location-based service becomes more volatile, it becomes harder to predict them both when implying an effort. Thus, the user becomes more reluctant to participate in the crowdsourcing. Therefore, the principal receives less utility and compensates the user with less reward.

In Fig. 3.3, we are planning to see the standard deviations of the two bonus items' influences on the compensation package. In Fig. 3.3a, d, we plot the optimal effort and principal's utility for both cases in the same figure, as the two bonus items have a symmetrical form in the expressions we derived in the previous section. Thus, as either σ_q and σ_p increases, the optimal effort and principal's utility have the same value. Similarly, we see that the values of the compensation package w in Fig. 3.3b, c are same when either of the two standard deviation varies.

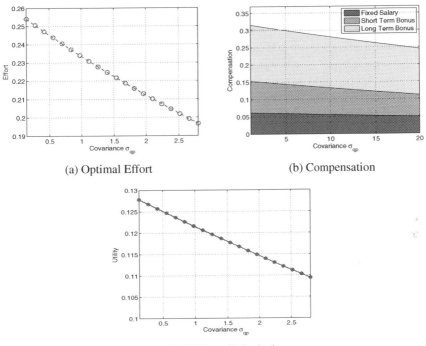

(a) Optimal Effort   (b) Compensation

(c) Utility of Principal

Fig. 3.2 Compensation package as the covariance σ_{qp} varies

From Fig. 3.3, we see that as each standard deviation (σ_q or σ_p) increases, the optimal effort a, compensation package w, and the principal's utility are decreasing. However, if we look inside the compensation package, the three bonus items show different properties. In Fig. 3.3b, we see that as the quality of location-based service's standard deviation σ_p increases, only the long-term bonus fp decreases with the compensation package w, while the fixed salary t and the short-term bonus sq are increasing. This situation is because as the principal's quality of location-based service becomes more volatile (σ_p increases), the user's benefit from the service decreases (fp becomes smaller), but the share from the fixed salary and transmitted data increases (t and sq increase).

Figure 3.3c shows similar properties as Fig. 3.3b. However, the two term bonus items sq and fp show the opposite behavior compared to the previous case. As σ_q increases, i.e., the quality of received data becomes more volatile, the user's compensation from the long-term bonus fp and the fixed salary t is increasing at the same time, while the short-term bonus sq in user's compensation package goes down.

From both Fig. 3.3b, c, we have learned that as the user's utility remains the same in all situations, the compensation package offered to the user will mostly rely on

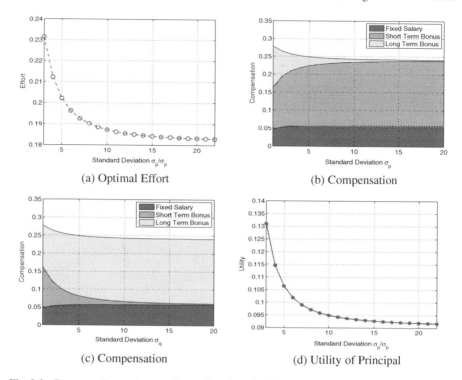

Fig. 3.3 Compensation package as the quality of received data or location-based service's standard deviation σ_q/σ_p varies

the part that is more stable, such as the fixed salary in both cases, while lowering the proportion of bonus from the less predictable part. By this mechanism, the risk of losing user's incentive in all kinds of situations can be canceled.

3.4.2   Compensation Package Comparison

In this subsection, we will propose another three incentive mechanisms as the comparisons with the compensation package in the previous section. For the incentive mechanism in the previous section, we have assumed that the compensation package includes three parts: the fixed salary t, the share of bonus based on current performance sq, and the share of bonus from long-term service fp. As we assume that the quality of received data and location-based service is the positively correlated, we give a notation of our mechanism as *Positive Covariance*. The other three mechanisms will be generally based on our current model, while different from each other in the construction of their compensation packages.

3.4.2.1 Independent

This mechanism is the special case of the *Positive Covariance*, as we assume that the two variables ε_q and ε_p are independently distributed, in which case, $\sigma_{qp} = 0$. Following the similar process in Sect. 3.3, the optimal effort a^* in this case is the same as the *Positive Covariance* in (3.14), and we obtain the expressions for f^*, s^*, and t^* as

$$s^* = \frac{\sigma_p^2}{\sigma_q^2 + \sigma_p^2 + \eta c \sigma_p^2 \sigma_q^2}, \tag{3.20}$$

$$f^* = \frac{\sigma_q^2}{\sigma_q^2 + \sigma_p^2 + \eta c \sigma_p^2 \sigma_q^2}, \tag{3.21}$$

$$t^* = \overline{w} + \frac{1}{2}\eta(s^2\sigma_q^2 + f^2\sigma_p^2) + \frac{1}{2}c\left(\frac{s+f}{c}\right)^2 - (s+f)\frac{s+f}{c}. \tag{3.22}$$

3.4.2.2 Single Bonus

In this case, we consider a compensation package which only includes the fixed salary t and short-term bonus sq, without the long-term bonus fp. We can regard this case as a company which does not directly provide the users with the location-based service based on their crowdsourcing activity. In this case, the optimal effort a^*, s^*, and t^* will be

$$a^* = \frac{s}{c}, \tag{3.23}$$

$$s^* = \frac{1}{1 + \eta c \sigma_q^2}, \tag{3.24}$$

$$t^* = \overline{w} + \frac{1}{2}\eta(s^2\sigma_q^2) + \frac{1}{2}c\left(\frac{s}{c}\right)^2 - \frac{s^2}{c}. \tag{3.25}$$

3.4.2.3 Opening Reward

The last mechanism is the simplest case in which the compensation package only contains a fixed salary t. We can regard this mechanism as a company which will offer each user an opening reward as the Karma we have mentioned in Sect. 3.1, while not caring about user's future effort. In this case, the optimal effort a^* and opening reward t^* have the form of

$$a^* = \frac{1}{c}, \tag{3.26}$$

Fig. 3.4 Principal's utility as the cost coefficient c varies

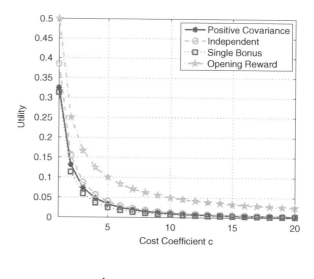

$$t^* = \overline{w} + \frac{1}{2c}. \tag{3.27}$$

3.4.2.4   Comparisons

In Fig. 3.4, we compare the principal's utility from the four incentive mechanisms as we vary the cost coefficient c. From the simulation results, we see that as the cost coefficient c increases, the principal's utility is decreasing as well. The reason for this phenomenon is because larger cost coefficient c means higher unit cost when implying an effort. Therefore, the user is less likely to participate in the crowdsourcing activity. With less data collected from the users, the principal's utility will decrease for sure. In addition, from Fig. 3.4, we see that the principal obtains the largest utility in the *Opening Reward* case, followed by the *Independent* one. The *Positive Covariance* mechanism proposed by us brings the third high utility to the principal, while the *Single Bonus* gives the least utility.

In Fig. 3.5, we analyze the impact of user's risk averse degree η on the principal's utility. As the principal's utility $V = a - t$ in the *Opening Reward* is independent of the risk averse degree η, we only compare the results from the other three mechanisms. We see that the principal's utility is decreasing as the user's risk averse degree η increases. This result is intuitive as a larger η means the user becomes more conservative and sensitive to risk, thus less likely to participate in. With less effort obtained from the user, the principal's utility will certainly decrease. From Fig. 3.5, we also obtain the similar ranking of the principal's utility as in the previous figure: The *Independent* case brings higher utility than the *Positive Covariance* one, and the *Single Bonus* one brings the smallest utility for the principal.

Fig. 3.5 Principal's utility as risk averse degree η varies

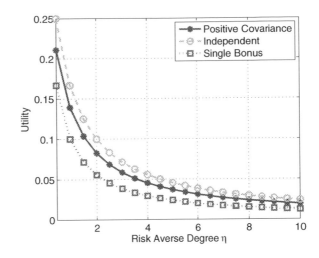

The reason for the performance ranking of the four mechanisms in Figs. 3.4 and 3.5 is as follows: The *Independent* mechanism is the ideal case of the *Positive Covariance*, in which less noise is occurred when predicting the outcome. Thus, higher utility is obtained than the general case—*Positive Covariance*, while the *Single Bonus* only rewards user with the current performance without providing the free location-based service. Thus, the users have fewer incentives to participate in. In return, less utility is obtained by the principal. For the result of the *Opening Reward* case, it seems unreasonable at the first sight. As it brings the principal higher utility than the other three mechanisms. While we notice that the *Opening Reward* is a "once-for-all" deal which does not provide continuous incentives for the users, i.e., after the users have fulfilled their duty and received the reward, they are more likely to stop participating in crowdsourcing.

3.5 Summary

In this chapter, we have proposed a performance-related compensation package for addressing the problem of incentivizing users to participate in crowdsourcing. We have included short- and long-term bonuses which relate to the quality of the received data from user and the quality of the location-based service provided by the principal, together with the fixed salary in the compensation package. In addition, we have proposed another three incentive mechanisms in special cases to compare with the proposed model. Through extensive simulations, we have given the analysis of the compensation results and shown that the users can be motivated by our proposed method. With this brief understanding of *moral hazard* model, we will move forward to more complicated models in the following chapters by extending the model to multi-user and multi-dimensional ones.

References

Angmin J, Valentino-Devries J (2011) Apple, google collect user data. Technic report. http://online.
    wsj.com/news/articles/SB10001424052748703983704576277101723453610
Bebchuk LA, Fried JM, Walker D (2002) Managerial power and rent extraction in the design of
    executive compensation. Univ Chicago Law Rev 69(2):751–846
Bolton P, Dewatripont M (2004) Contract theory. The MIT Press, Cambridge, MA
Duan L, Kubo T, Sugiyama K, Huang J, Hasegawa T, Walrand J (2012) Incentive mechanisms
    for smartphone collaboration in data acquisition and distributed computing. In: The 31st IEEE
    international conference on computer communications (INFOCOM), Orlando, FL
Gao L, Iosifidis G, Huang J, Tassiulas L (2014) Hybrid data pricing for network-assisted user-
    provided connectivity. In: INFOCOM, 2014 Proceedings IEEE, Toronto, Canada
Karma (2012) Meet karma. Technic report. https://yourkarma.com/
Luo T, Tan H, Xia L (2014) Profit-maximizing incentive for participatory sensing. In: INFOCOM,
    2014 Proceedings IEEE, Toronto, Canada
Tirole BHJ (1993) Market liquidity and performance monitoring. J Politi Econ 101(4):678–709
Werin L, Wijkander H (1992) Contract economics. Blackwell Publishers, Oxford, UK
Zhao D, Li X, Ma H (2014) How to crowdsource tasks truthfully without sacrificing utility: online
    incentive mechanisms with budget constraint. In: INFOCOM, 2014 Proceedings IEEE, Toronto,
    Canada

Chapter 4
Tournament-Based Incentive Mechanism Designs for Mobile Crowdsourcing

4.1 Introduction

In the previous chapter, the *moral hazard* problem in mobile crowdsourcing is between one employer (the principal) and one employee (the user). In this chapter, we will consider a more complicated model with multiple number of users, which better meets the requirement of reality. As a large number of users are needed to support the crowdsourcing activity. Nowadays, people can access various sophisticated location-based services (e.g., Google Maps with traffic information) using their smartphones through wireless access networks. With the drastic growth in the location-based service market, as well as the rapid development of big data technologies, more data as well as user participation are required to support more sophisticated services. There are mobile applications available that can detect Wi-fi hot spot within a certain distance of the user's current location. Smartphone users help collect the Wi-fi hot spot information that includes the location and router name for the service provider which is denoted as principal hereafter. However, when participating in such crowdsourcing, users consume their resources such as battery and computing capacity (Zhao et al. 2014). Therefore, many users hesitate to participate which is a major impediment to the growth of mobile crowdsourcing (Angmin and Valentino-Devries 2011). Thus, incentive mechanism designs are in critical need to motivate the users to participate.

In the literature, it has already been noticed that there is an urgent need to alleviate the conflict by introducing incentive mechanism for users. Inspired by the effort-based reward from the labor market, several works have been proposed to address this problem by providing users with the reward that is consistent with their performance. Examples are the works in Luo et al. (2014) and Duan et al. (2012), as well as one of our previous works (Zhang et al. 2015). The previous mentioned works capture the fundamental aspect of providing necessary incentive for user to participate in crowdsourcing. Yet, they mainly assume that the principal employs only one user and rewards it on the basis of the absolute performance.

However, when rewarding users based on the absolute performance, the principal has a strong incentive to cheat by claiming that users had poor performances that

© Springer International Publishing AG 2017
Y. Zhang and Z. Han, *Contract Theory for Wireless Networks*,
Wireless Networks, DOI 10.1007/978-3-319-53288-2_4

deserve low rewards, so that the principal can pay less (Bolton and Dewatripont 2004). Another example is that when there is a positive mean measurement error at user's performance, every user's performance will result in an increase at the principal's observation. We name this case that affects both sides as *common shock*, which can be either positive or negative to user performance and reward. If both users and principal are aware of this *common shock*, we can regard the trading between them as trading with *full information*. However, in the general case, this *common shock* is unobservable to either side or both sides. While incentive mechanism based on the absolute performance can be easily affected, the tournament design can filter out this *common shock* problem.

One obvious advantage of rank-order tournament over absolute performance rewards is that ordinal ranking is easy to measure and hard to manipulate (Bolton and Dewatripont 2004). In a tournament, the principal has to offer the fixed amount of rewards no matter who wins. In this chapter, we will propose a multi-user design that rewards users' performance in crowdsourcing by a tournament reward structure based on the rank order. A brief illustration of crowdsourcing tournament rewarding mechanism is shown in Fig. 4.1. After obtaining the data from the users, the principal will generate an ascending list regarding user's performance. Here, user 1 achieves the highest performance and will be rewarded the highest amount reward 4, while user 2 performs worst with the smallest amount of reward 1.

The main contributions to this chapter are as follows: First, we solve the incentive problem in crowdsourcing by offering user reward based on their performance.

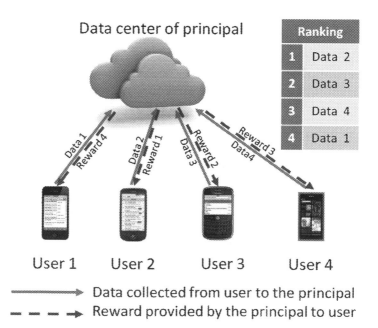

Fig. 4.1 Crowdsourcing incentive mechanism by tournament

Second, we consider a tournament structure incentive mechanism that rewards users by their rank orders which can overcome the *common shock* problem. Third, in the simulation part, we introduce another well-known tournament mechanism for comparison purposes in order to demonstrate the effectiveness of tournament mechanisms to improve the principal's utility. The proposed mechanisms allow the principal to successfully maximize the utilities and the users to obtain continuous incentives to participate in mobile crowdsourcing.

The remainder of this chapter is organized as follows. First, we will introduce the network model in Sect. 4.2. Then, the design of tournament is described in Sect. 4.3, in which we also give the analysis of the optimal contract with full information. The performance evaluation is conducted in Sect. 4.4. Finally, summaries are given in Sect. 4.5.

4.2 System Model

We refer to the model in Lazear and Rosen (1981) and consider a crowdsourcing network in which one risk-neutral principal employs a fixed group of identical risk averse users, $i = 1, \ldots, n$, to collect data. The principal rewards users based on their relative performances which can be referred to the quality of the received data (e.g., quantity, correctness, and importance). In a n-user tournament, the users' performances are sorted in an ascending order, and fixed prizes $(W_1, W_2, \ldots, W_n)$ are rewarded. We use the numbering conventional in the study of order statistics: "First place" is the lowest performance, and W_1 is the prize received by the user with the lowest performance.

4.2.1 Common Shock Problem

When users help to collect data for the principal, the user exerts an effort a. Note that the user's effort a is a hidden information, since the principal can only observe the performance level q of the users, i.e., the quality of the received data. Therefore, the performance of user i, q_i, depends stochastically on the user's effort level, a_i. In particular,

$$q_i = z_i + \varepsilon, \tag{4.1}$$

where ε is a random variable representing the *common shock* that affects all of the users and z_i is a random variable whose distribution depends on a_i. Due to the *common shock*, such as the measurement error of the principal as mentioned previously, the quality of received data q_i cannot reflect the user's actual performance or effort exactly. Therefore, the performance of the user is a noisy signal of its effort.

Let G denote the distribution function for the *common shock* (μ, σ^2), where σ^2 is the variance. We assume that ε has zero mean when no *common shock* presents:

$$\int \varepsilon dG(\mu, \sigma^2) = 0. \tag{4.2}$$

By this assumption, regardless of its assessment of ε, every user believes that its performance and that of every other user have the same mean if they take the same effort.

4.2.2 Rank-Order Statistic

Let $F(z_i; a_i)$ denote the cumulative distribution function (CDF) for z_i, given a_i. $F(z_i; a_i)$ has a continuous probability distribution function (PDF) $f(z_i; a_i)$ which is positive everywhere and continuously differentiable in a_i. Since the users are identical ex ante, F does not depend on i. The value of z_i is not known to the user until its choice of a_i is made. We assume that z_i and (μ, σ^2) are independent, since the term z_i is independently and identically distributed for every common value of a_i and q_i.

Assume that the principal observes only the performance levels of the users, $q = (q_1, q_2, \ldots, q_n)$, but cannot directly observe the users' effort levels. Under the tournament, user i's reward depends only on the rank order of q_i in q, instead of the performance level q_i. Since each user's performance is given by $q_i = z_i + \varepsilon$, we can easily obtain $z_i \geq z_j$ from $q_i \geq q_j$. That is, the rank order of the performances depends only on z_i and not ε. Therefore, the realization of (μ, σ^2) does not affect the game played by the users, and the equilibrium effort level will be independent of σ^2. Hence, we can analyze the game in terms of just z_i. In a n-user tournament, user i wins prize W_j if and only if z_i is the jth-order statistic of $(z_1, \ldots, z_n)$. The density function $\phi_{jn}(z; a)$ for the jth-order statistic in a sample of size n drawn from the distribution $F(z; a)$ is Green and Stokey (1983)

$$\frac{(n-1)!}{(n-j)!(j-1)!} f(z; a) F^{j-1}(z; a)[1 - F(z; a)]^{n-j}. \tag{4.3}$$

This density function denotes that the user i's performance outperforms $j-1$ number of users and falls behind $n-j$ number of users.

4.2.3 Utility of the Users

The realized performance of each user then is a stochastic function of its effort and the value of the *common shock*. Here, we consider the user's reward from the principal's prize in terms of utility. It is also convenient to think of the cost of exerting effort in terms of utility. The preferences of each user i over the prize, W_i, and the exerted effort, a_i, are represented by the utility function

$$U_t(W_i, a_i) = u(W_i) - \gamma(a_i), \tag{4.4}$$
$$W_i \geq 0, \quad a_i \geq 0, \quad i = 1, \ldots, n,$$

where u is a strictly increasing and strictly concave function of W_i, and γ is strictly increasing and strictly convex with a_i. The user's utility is obtained from the prize minus the exerting effort.

For convenience, the principal can construct the user's reward function in terms of utility $w = (w_1, w_2, \ldots, w_n)$ by defining $w_i = u(W_i)$, $\forall i$. We have the user's expected utility as the expected value of rewards minus the cost,

$$U_t(w, a) = \sum_{j=1}^{n} w_j P(\text{rank} = j) - \gamma(a), \tag{4.5}$$

where $P(\text{rank} = j)$ is the probability that the user is in the jth place among all n users at the measured performance level $q = z + \varepsilon$. Given the density function $\phi_{jn}(z; a)$, the probability can be obtained by an integration of the density function $\phi_{jn}(z; a)$. Thus, the user's utility function can be written as

$$U_t(w, a) = \sum_{j=1}^{n} w_j \int \phi_{jn}(z; a) dz - \gamma(a). \tag{4.6}$$

In the symmetric equilibrium, all users spend the same amount of effort $\bar{a}$ and expect an equal probability $1/n$ of reaching any of the n ranks. Given the effort choice of $\bar{a}$, we can derive the users' expected utility from (4.6) as

$$U_t(w, \bar{a}) = \frac{1}{n} \sum_{j=1}^{n} w_j - \gamma(\bar{a}). \tag{4.7}$$

4.2.4 Utility of the Principal

The principal's problem is to design a reward structure for the n-users. We assume that the principal is constrained to offer a fixed minimum level of expected utility to each user, so that we can judge the relative performance of tournaments by examining the expected utility of the principal. The risk-neutral principal's objective is to maximize the summation of all the users' performances minus the total prizes to the users:

$$V_t(W, a) = E\left[\sum_{i=1}^{n} (q_i - W_i)\right]. \tag{4.8}$$

Given that the performance q follows a conditional distribution $f(q - \varepsilon, a)$ and under a *common shock*, the principal's expected utility can be written as:

$$V_t(w, a) = \int \int q f(q - \varepsilon, a) dG(\varepsilon, \sigma^2) dq - \sum_{j=1}^{n} W_j, \tag{4.9}$$

$$= \int z f(z, a) dz - \sum_{j=1}^{n} W_j, \tag{4.10}$$

where (4.9) is result from our previous conclusion that z is independent from the common shock (ε, σ^2), and thus, we can simply replace q with z.

4.3 Problem Formulation

4.3.1 Optimization Problem

Given the number of users n that participate in this crowdsourcing, the principal's problem is to design $(w, \bar{a})$ to maximize (4.9) subject to the two constraints that $\bar{a}$ is an optimal decision rule for the user given w and that the expected utility of the user is at least $\bar{u}$, i.e.,

$$\max_{(w, \bar{a})} \int z f(z, a) dz - \sum_{j=1}^{n} W_j, \tag{4.11}$$

s.t.

$$(a) \quad \bar{a} = \arg\max_a \sum_{j=1}^{n} w_j \int \phi_{jn}(z; a) dz - \gamma(a),$$

$$(b) \quad \frac{1}{n} \sum_{j=1}^{n} w_j - \gamma(\bar{a}) \geq \bar{u}.$$

where (**a**) is the incentive compatible (IC) constraint; it represents that given any reward structure, the problem facing each user is to choose a level of effort that maximizes own utility. We can solve the optimal effort by taking the first derivative of the IC constraint, which is given by

$$\sum_{j=1}^{n} w_j \frac{\partial P(\text{rank} = j)}{\partial a} - \gamma'(a) = 0. \tag{4.12}$$

where (**b**) is the individual rationality (IR) constraint; it provides the necessary incentive for users to participate. We must have the utility no less than the reservation utility when a user is not taking any effort ($a = 0$). Here, we denote $S_t(n)$ as the set of feasible n-user tournaments that satisfy the IC and IR constraints. The set of feasible tournaments is always non-empty, since it always contains the "no incentive" tournament, $[(\bar{u}, \bar{u}, \ldots, \bar{u}), 0] \in S_t(n)$, for all n. The utility per user to the principal under this tournament is $\bar{V}$.

From the problem formulation, we see that the optimal tournament depends on the number of users n and the distribution function F, but not on the distribution function G. In other words, tournament approach is robust against the lack of information or lack of agreement about G.

4.3.2 Tournament Design

To obtain the tournament, we can derive from the optimal contract that rewards user based on the absolute performance with full information. First, we will formulate the optimal contract problem with full information. Then, we will show that we can design the tournament by step functions to approximate the optimal contract.

4.3.2.1 Optimal Contract Under Full Information

In the optimal contract, the principal rewards users based on the absolute performance. We define the reward $R(q)$ as a linear and increasing function of q. Thus, the utility user obtained from the reward is $u(R(q))$ and denoted as $v(q)$ for simplicity. The contract the principal offered to user is (v, A), where A is the effort. In this full information case, G is given by $\varepsilon = 0$ with probability 1; i.e., the principal knows ε.

Thus, the user i's utility under contract is represented by

$$U_c(v_i, a_i) = v(q_i) - \gamma(a_i), \tag{4.13}$$
$$q_i \geq 0, \quad a_i > 0, \quad i = 1, \ldots, n,$$

where v is also a strictly increasing and concave function as u. As we can see, $v(q_i)$ is a piecewise continuous utility function which related to the quantity of q_i instead of its rank. As noted above, $F(z; a)$ denotes the conditional distribution function for z given a, and $f(z; a)$ is the continuous density function of $F(z; a)$. As $\varepsilon = 0$ with probability 1, we can rewrite the user's expected utility function as

$$U_c(v, a) = \int v(z) f(z; a) dz - \gamma(a), \tag{4.14}$$

which is positive everywhere and continuously differentiable in a.

Followed by user's expected utility function in contract, the principal's expected utility can be written as

$$V_c(v, a) = E\left[\sum_{i=1}^{n} (q_i - R(q_i))\right],$$ (4.15)

where $\gamma[v(q_i)]$ is the cost utility function which is also a strictly increasing and strictly convex function of the utility provided to user, as in the tournament. Similarly, the expected utility of the principal of the contract (v, a) is

$$V_c(v, a) = \int \{z - R(z)\} f(z; a) dz.$$ (4.16)

With the user and principal's utility functions, we can formulate the contract which rewards user by their absolute performance as

$$\max_{(v,A)} \int \{z - R(z)\} f(z; a) dz,$$ (4.17)

$$s.t.$$

$$(a) \quad A = \arg\max_a \int v(z) f(z; a) dz - \gamma(a),$$

$$(b) \quad \int v(z) f(z; A) dz - \gamma(A) \geq \bar{u}.$$

Similar to the tournament, (a) is the IC constraint and (b) is the IR constraint. The principal's problem is to choose (v, A) to maximize its expected utility subject to the two constraints that A is the optimal decision rule for the user given v and that the expected utility of the user is at least $\bar{u}$. Here, we denote S_c as the set of feasible contracts that satisfy the IC and IR constraints. From Green and Stokey (1983), the piecewise continuous utility function and the user's optimal effort can be approximated arbitrarily closely by a step function, if there are enough steps.

4.3.2.2 Tournament by Approximation

Next, we will show that given a feasible contract $(v, A) \in S_c$, we can approximate the optimal contract by constructing a sequence of contracts $(w_{ni}, \bar{a}_n)$, where w_{ni} is a step function with n steps and $\bar{a}_n$ is a constant function.

The first thing we need to do is to approximate the continuous utility function $v(z)$ by a step function. We notice that the probability that a user achieves a specific rank is equal to the probability that the user's performance level falls into a corresponding interval of the CDF. Thus, given a specific rank, we can find the effort value q_{ni} by the inverse CDF of $F(q_{ni}; A) = i/(n + 1)$ (Green and Stokey 1983). Then, we can define $\hat{w}_{ni}$ by

$$\hat{w}_{ni} = v(q_{ni}), \quad i = 1, \ldots, n.$$ (4.18)

Thus, we can replace the w_j in (4.11) with this approximation $\hat{w}_{ni}$. The optimal effort under tournament can be solved by

$$\bar{a}_n = \arg\max_a \sum_{i=1}^{n} \hat{w}_{ni} \int \phi_{in}(z; A)dz - \gamma(A). \qquad (4.19)$$

Again, we calculate the error term $\bar{e}_n$ in this tournament design and have

$$\bar{e}_n = \bar{u} + \gamma(\bar{a}_n) - \frac{1}{n} \sum_{i=1}^{n} \hat{w}_{ni}. \qquad (4.20)$$

Finally, the utility in tournament is obtained by adding up the approximated $\hat{w}_{ni}$ and error $\bar{e}_n$:

$$w_{ni} = \hat{w}_{ni} + \bar{e}_n, \quad i = 1, \ldots, n. \qquad (4.21)$$

By now, we have the tournament $(w_{ni}, \bar{a}_n)$ that is close to the optimal contract with full information.

Each of these step function contracts can be approximated arbitrarily close by a tournament with a sufficiently large number of users. Hence, the principal's expected utility is approximately unchanged. Moreover, the tournament's efficiency is unaffected by changes in G (the distribution of ε and the user's information about ε), so that the same tournament's utility remains arbitrarily close to the full information utility for any G as well as if the users can observe ε directly.

4.4 Simulation Results and Analysis

In this section, we will give numerical simulations to illustrate our results. First, we will give the specific form of the utility and cost functions we have defined in the system model. Then, we will show the tournament we obtained by the step function. Finally, we will analyze the system performance by varying different parameters and do a comparison with other incentive mechanisms.

4.4.1 Simulation Setup

We assume that the conditional distribution follows the logistic distribution as Kalra and Shi (2001). The logistic distribution is a symmetric and bell-shaped distribution, like the frequently used normal distribution. The PDF of a logistical distribution is

$$f(z; a) = \frac{\exp(-\frac{z-a}{\beta})}{\beta[1 + \exp(-\frac{z-a}{\beta})]^2},$$

(4.22)

and the CDF is

$$F(z; a) = \frac{1}{1 + \exp(-\frac{z-a}{\beta})},$$

(4.23)

where β is the coefficient related to the variance of logistic distribution, which is $\pi^2 \beta^2 / 3$. As β is positively correlated with the variance, we will use β to denote the variance in the sequel. With the PDF and CDF of logistic distribution, we can derive the partial derivative of the probability for the jth-order statistic with respect to effort a, as

$$\frac{\partial P(\text{rank} = j)}{\partial a} = \frac{2j - n - 1}{\beta[n(n + 1)]}.$$

(4.24)

Proof To obtain the simplified form of $\frac{\partial P(\text{rank}=j)}{\partial a}$ which is first derivative of the the probability of ranking j, we can make use of convenient form of the logistic distribution by the following procedures. First, we take the first derivative of the probability of ranking j with respect to effort result z and rewrite $\frac{\partial P(\text{rank}=j)}{\partial a}$ as

$$\frac{\partial P(\text{rank} = j)}{\partial a}$$

(4.25)

$$= \int \frac{(n-1)!}{(n-j)!(j-1)!} \{(n-j)[1 - F(z; a)]^{n-j}(-f(z; a))F^{j-1}(z; a)f(z; a)$$

$$+ [1 - F(z; a)]^{n-j}(j-1)F^{j-2}(z; a)f^2(z; a)\}dz,$$

$$= \int \frac{(n-1)!}{(n-j)!(j-1)!}[1 - F(z; a)]^{n-j-1}F^{j-2}(z; a)f^2(z; a)[(j-1)$$

$$- (n-1)F(z; a)]dz,$$

$$= \frac{(n-1)!}{(n-j)!(j-1)!}\{(j-1)\int[1 - F(z; a)]^{n-j-1}F^{j-2}(z; a)f^2(z; a)dz$$

$$- (n-1)\int[1 - F(z; a)]^{n-j-1}F^{j-1}(z; a)f^2(z; a)dz\}.$$

Taking the specific form of the logistic distribution into $\psi(j)$, we have

$$\frac{\partial P(\text{rank} = j)}{\partial a}$$

(4.26)

$$= \frac{(n-1)!}{(n-j)!(j-1)!}$$

$$\{(j-1)\int \frac{\exp(-\frac{x}{\beta})^{n-j-1+2}}{[1 + \exp(-\frac{x}{\beta})]^{n-j-1+j-2+4}}dz$$

$$-(n-1)\int \frac{\exp(-\frac{x}{\beta})^{n-j-1+2}}{[1+\exp(-\frac{x}{\beta})]^{n-j-1+j-1+4}}dz\},$$

$$=\frac{(n-1)!}{(n-j)!(j-1)!}\{(j-1)\int \frac{\exp(-\frac{x}{\beta})^{n-j+1}}{[1+\exp(-\frac{x}{\beta})]^{(n-1)+2}}dz$$

$$-(n-1)\int \frac{\exp(-\frac{x}{\beta})^{n-j+1}}{[1+\exp(-\frac{x}{\beta})]^{n+2}}dz\},$$

For the logistic distribution, there is a property that

$$\int \frac{\exp(-\frac{x}{\beta})^k}{[1+\exp(-\frac{x}{\beta})]^{n+2}}dx = \frac{(k-1)!(n-k+1)!}{(n+1)!}\beta, \qquad (4.27)$$

when $k \geq 2$. With this property, we can simplify an integration to a fraction. Thus, we are able to simplify $\psi(j)$ as follows:

$$\frac{\partial P(\text{rank}=j)}{\partial a} = \qquad\qquad\qquad\qquad\qquad\qquad\qquad\qquad (4.28)$$

$$=\frac{(n-1)!}{(n-j)!(j-1)!}\{\frac{(j-1)}{\beta}\frac{(n-j+1-1)![n-1-(n-j+1)+1]!}{(n-1+1)!}\beta$$

$$-\frac{(n-1)}{\beta}\frac{(n-j+1-1)![n-(n-j+1)+1]!}{(n+1)!}\beta\},$$

$$=\frac{(n-1)!}{(n-j)!(j-1)!}\{\frac{(j-1)}{\beta}\frac{(n-j)!(j-1)!}{n!}\beta - \frac{(n-1)}{\beta}\frac{(n-j)!j!}{(n+1)!}\beta\},$$

$$=\frac{2j-n-1}{\beta[n(n+1)]}.$$

Now, we obtain the simplified form of the partial derivative $\frac{\partial P(\text{rank}=j)}{\partial a}$. □

According to (4.12), we must have $2j - n - 1 > 0$. Thus, the maximum number of reward recipients will not be more than half of the participate users. The reward recipients should be the users whose rank is higher than $(n+1)/2$, while the users whose rank is lower than $(n+1)/2$ will only receive a zero reward.

In the system model, we have defined the evaluation function u as a concave function. Here, we set up the evaluation function u in a form of power function as

$$u(W) = \frac{W^\rho}{\rho}, \qquad (4.29)$$

where ρ is the power coefficient and $0 < \rho < 1$. Here, we further define the user's risk averse degree as

$$\eta = -\frac{u'}{u''} = \frac{1-\rho}{W}. \qquad (4.30)$$

Under the same amount of reward, the larger the ρ and η, the more conservative and sensitive is user toward risk, and vice versa. ρ and η are negatively correlated with each other. When $\rho = 1$, the user is risk neutral. For simplicity, we define the reward function as $R(q) = q$. Thus, the utility function in the optimal contract case becomes

$$v(q) = u[R(q)] = u(q) = \frac{q^\rho}{\rho}. \tag{4.31}$$

Furthermore, we have defined the cost function in the system model as a convex function. Thus, we set up the cost function γ in a quadratic form as

$$\gamma(a) = \frac{1}{2}a^2. \tag{4.32}$$

We assume that the reservation utility, when the user does not participate in the crowdsourcing, is $\bar{u} = 0$.

4.4.2   Reward by Tournament

In Fig. 4.2, we follow the steps in Sect. 4.3 to approximate the optimal contract by tournament with 19 users participate in, with x-axis representing the rank of the users in an ascending order. As we can see, the reward obtained by the tournament is close to

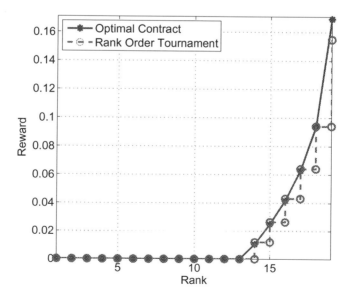

Fig. 4.2 Approximation of optimal contract by tournament

the reward from the optimal contract with full information. If we increase the number of users to infinity, the tournament can approximate the optimal contract arbitrarily close. In addition, we see that only users rank is larger or equal to 14 received a positive reward, which is consistent with our conclusion previously that no more than half of the users should be rewarded. Another observation from Fig. 4.2 is that the higher the user rank, the larger the spread is, that is, $W_j - W_{j-1} < W_{j+1} - W_j$. This result is due to the power function form of the evaluation function u. If we change the evaluation function u to a log function, the spread will be the same for all ranks. While if the evaluation function u follows the exponential form, the spread will become smaller for higher ranks.

4.4.3 Comparison

In this part, we are going to analyze user and principal's utilities by varying different parameters in the tournament. In the tournament we have proposed, there are many winners and the amount of reward is based on the relative rank achieved, with larger amounts rewarded to higher ranks. We refer to this as the rank-order tournament (ROT). We will compare the results from the ROT with that from the optimal contract with full information and another special cases of ROT: the Multiple-Winners (MW) in which several top winners share the reward equally and the optimal number of winners can be determined from (4.12).

From Figs. 4.3a–4.5a, we show the utility per user when varying different parameters, and from Figs. 4.3b–4.5b, we show the utility of the principal. The figures show that the factors impacting the design of the contest include the number of users for whom the contest is conducted, the degree of performance uncertainty in the environment (i.e., the strength of the relation between effort and performance realized), and the user's risk averse degree toward the crowdsourcing activity.

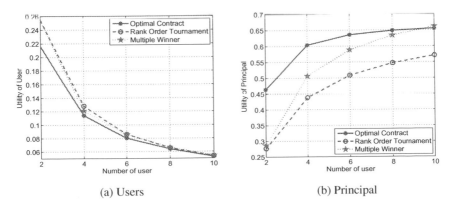

(a) Users   (b) Principal

Fig. 4.3 The utility of the user and principal as the number of users vary

4.4.3.1 Number of Users

When the number of users n increases, the marginal change in the probability of achieving any rank decreases. Consequently, with increases in the pool of players, the user will be less likely to induce higher effort levels and less incentive to participate. Thus, we see the user's utility in Fig. 4.3a decreases with the increase of n. However, with more users participating in the crowdsourcing, even though the effort exerted from each user decreases, the summation of the data collected with more number of users increases. As a result, the principal's utility increases as we see from Fig. 4.3b.

4.4.3.2 Variance

The variance β denotes the relation between effort levels exerted by the user and the performance observed by the principal. As β increases, it indicates a weaker relation between effort levels and the expected rank achieved. As a result, the users are likely to exert lower levels of effort for the increase in uncertainty and thus a lower cost of participation. While the optimal contract and tournament designs are independent of the uncertainty, greater uncertainty makes users more likely to get enough incentives to participate. As we see from Fig. 4.4a, the user's utility is increasing as the variance increases. With the decrease in optimal effort, less data is obtained from the user, and the principal's utility will certainly decrease. Therefore, Fig. 4.4b indicates that the principal's utility is decreasing as the variance increases.

4.4.3.3 Risk Averse Degree

From the definition of risk averse degree, we see that when η increases, users become more conservative and sensitive to risk, thus less likely to participate in. With less

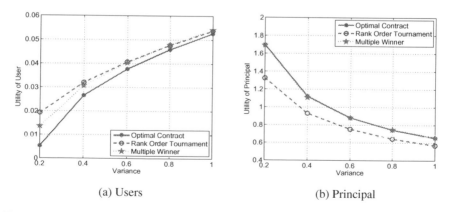

(a) Users   (b) Principal

Fig. 4.4 The utility of the user and principal as the variance vary

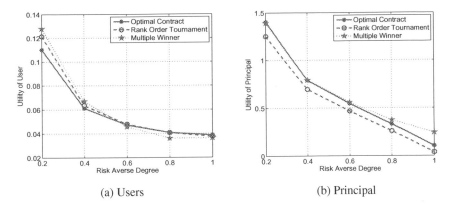

Fig. 4.5 The utility of the user and principal as the risk averse degree vary

effort obtained from the user, the principal's utility will certainly decrease. Thus, we see from Fig. 4.5a, b that the user and principal's utilities decrease with the risk averse degree η.

Overall, we see that the optimal contract serves as the upper bound of the principal's utility, and the lower bound of the user's utility for the other two tournament mechanisms in most of the cases. This is intuitive since the optimal contract solves the optimal contract based on the absolute performance. While in tournament, we only have a limited number of users in the simulation. Thus, tournaments lose accuracy during the approximation. The optimal contract provides the principal with a maximum utility while extracting as much utility from the users as possible.

From Figs. 4.3 and 4.5, we also see that the MW outperforms ROT in many cases. In addition, MW outperforms both the optimal contract and ROT when users are risk neutral as shown in Fig. 4.5a, b. The reasons for both results can be inspired from the conclusions drawn in Kalra and Shi (2001). First, when the number of participating users is small, MW is a better mechanism rather than ROT and is easier to implement. Here we only consider no more than 10 participating users due to the computation capacity of our computer. With such a small group of users in our simulation, we see MW outperforms ROT in all simulation results. Second, when users are risk neutral, it is optimal to give the entire reward to the highest rank user, which is a special case of MW, rather than offering contract with positive spread in ROT and optimal contract.

4.5 Summary

In this chapter, by using the multi-user *moral hazard* model, we have investigated the problem of providing incentives for users to participate in mobile crowdsourcing by applying the rank-order tournament as the incentive mechanism. We have solved the rank-order tournament by approximating the absolute performance-based optimal contract with full information using step functions. Finally, we use the numerical results to show the tournament design and compare the user's and principal's utilities under the optimal contract and different tournament mechanisms. We have shown that by using the tournament, the principal successfully maximizes utilities regardless of *common shock*. The principal's utility benefits from large number of users, but deteriorates into weaker relationship between exerted and observed effort levels, and higher risk aversion of users. After discussing this multi-user *moral hazard* model, we will give another extension of the fundamental case, in which a multi-dimension reward will be given to evaluate an employee's performance.

References

Angmin J, Valentino-Devries J (2011) Apple, google collect user data. Technic report. http://online. wsj.com/news/articles/SB10001424052748703983704576277101723453610

Bolton P, Dewatripont M (2004) Contract theory. The MIT Press, Cambridge

Duan L, Kubo T, Sugiyama K, Huang J, Hasegawa T, Walrand J (2012) Incentive mechanisms for smartphone collaboration in data acquisition and distributed computing. In: INFOCOM, Proceedings IEEE, Orlando, FL

Green JR, Stokey NL (1983) A comparison of tournaments and contracts. J Polit Econ 91(3):349–364

Kalra A, Shi M (2001) Designing optimal sales contests: a theoretical perspective. Mark Sci 20(2):170–193

Lazear EP, Rosen S (1981) Rank-order tournaments as optimum labor contracts. J Polit Econ 89(5):841–864

Luo T, Tan H, Xia L (2014) Profit-maximizing incentive for participatory sensing. In: INFOCOM, Proceedings IEEE, Toronto, Canada

Zhang Y, Gu Y, Liu L, Pan M, Dawy Z, Han Z (2015) Incentive mechanism in crowdsourcing with moral hazard. In: TIEEE wireless communications and networking conference (WCNC), New Orleans, LA

Zhao D, Li X, Ma H (2014) How to crowdsource tasks truthfully without sacrificing utility: online incentive mechanisms with budget constraint. In: INFOCOM, Proceedings IEEE, Toronto, Canada

Chapter 5
Multi-dimensional Payment Plan in Fog Computing with Moral Hazard

5.1 Introduction

In the previous two chapters, we have discussed two applications by using different *moral hazard* models. In this chapter, a multi-dimension reward model will be given to design-efficient payment plan in wireless networks. The rapid developments of cloud computing have brought a centralized solution to application developers and content providers. Despite its widely known conveniences and advantages, cloud computing also suffers from certain limitations such as high latency and delay due to long distance between end users and servers (Bonomi et al. 2012). The emerging trends in networking such as large distributed sensor networks, industrial automation, and high-speed transportation need location-dependent fast processing and cannot be satisfied by the current service form by cloud computing (Barbarossa et al. 2014).

With the motivation of placing the services as close as possible to end users, researchers have proposed a new cloud system called fog computing. In this model, fog nodes (FNs) such as end user devices, access points, edge routers and switches are deployed at or very close to the edge of network, and with functionalities such as converged computing, processing, management, networking, storage, physical, and cyber security (Patel et al. 2014). Technically, fog computing is similar to cloud computing in the sense that both are made up of virtual systems providing the flexibility of on-demand provisioning of compute, storage, and network resources. However, fog computing has several advantages over cloud computing in the sense of a significant reduction in data movement across the network resulting in reduced congestion, cost and latency, and elimination of bottlenecks resulting from centralized computing systems, improved security of encrypted data as it stays closer to the end user reducing exposure to hostile elements, and improved scalability arising from virtualized systems (Cisco 2015b). By opening the access to fog computing nodes, service providers (SPs) can rapidly deploy certain applications and services to improve the quality of service (QoS) toward end users. This environment can also create a new value chain comprising NOs, InPs, SPs, and end users.

© Springer International Publishing AG 2017
Y. Zhang and Z. Han, *Contract Theory for Wireless Networks*,
Wireless Networks, DOI 10.1007/978-3-319-53288-2_5

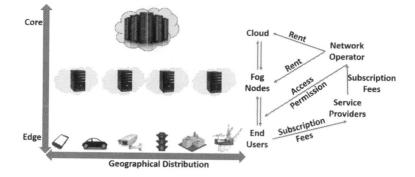

Fig. 5.1 An illustration of fog computing system

Referring to one brief model of fog computing in Fig. 5.1, there are variety of end users from the areas such as smart grid, industry, vehicular networks, transportation system, and public safety department that require real-time computing services. By subscribing to specific SPs whom subscribed to the network operator (NO) to obtain access to physical resources, end users are allowed to access computing resources in both the fog and the cloud, with fog close to end users while cloud locates far away. Within this network, end users directly communicate with FN for real-time control and analytic, while the FNs only send periodic data summaries to the cloud for further aggregation and procession. Usually, the cloud and fog are managed by the NO, who rents the cloud center and fog nodes from infrastructure providers (InPs).

The NO is aiming at maximizing its revenue by efficiently managing and coordinating the computing resources in both fog and cloud. To fully utilize the fog computing with minimal rent while ensuring the FNs receive nonnegative revenue, appropriate payment plan is needed to enable the NO and FNs to play complementary roles within their respective business models, and allow all players to benefit from greater cooperation. Inspired by the effort-based reward from the labor market where employers pays its employees based on their work load, we propose a payment plan in fog computing such that the FNs receive their rental in accordance with the quantities of computing resources and the quality of service (QoS) they provide to the NO. The computing resources include the transmission bandwidth, power, computation capability (CPU speed), storage size, and a FN's proximity toward end users. Meanwhile, the QoS can be referred to latency and delay during data transmission and processing, as well as security (Sardellitti et al. 2015).

Based on this motivation, we aim at offering a contract that considers different aspects of the computing resources provided by FNs to end users and assigns different payment weights in order to maximize the revenue of NO. Fortunately, the moral hazard model of contract theory provides us a useful tool to design such a payment plan that can solve the NO's revenue maximization problems in fog computing when the FN's performance in QoS is affected by multiple aspects (Bolton and Dewatripont 2004). From the NO's perspective, it "employs" the FNs to perform computing tasks and offers them QoS-consistent payment by multi-dimension measurements. Inside this value chain, the NO tries to guarantee the fog computing QoS with minimal payment, while ensuring the FNs have necessary incentives to cooperate. Thus, to maximize the NO's revenue, the NO needs to find an optimal payment plan that can properly pay the infrastructure rent to FNs (Werin and Wijkander 1992).

The main contributions to this chapter are summarized as follows. First, we propose a QoS-consistent contract that considers the quantities of resources provided by FNs with multiple aspects. The contract characterizes the general situation in the real world and provides a comprehensive payment plan to the FNs for using those resources. Second, we formulate NO's revenue maximization problem, as well as provide the FNs with the necessary incentive to cooperate in fog computing. Third, through simulations, we reveal different parameter's impacts on the optimal payment plan and compare the NO utility under six different payment plans. Our results show that by using the proposed payment plan, the NO successfully maximizes the utilities and the FNs obtain the continuous incentives to participate in the fog computing.

The remainder of this chapter is organized as follows. First, we will introduce the network model in Sect. 5.2. Then, the problem formulation is described in Sect. 5.3. The performance evaluation is conducted in Sect. 5.4. Finally, Sect. 5.5 summarizes this chapter.

5.2 System Model

In this section, we consider a monopoly market with one NO trading with one FN. The NO–FN mutual benefit model is introduced first by constructing the multi-dimension payment plan offered by the NO. Then, we will give the utility functions of both the FN and the NO before proceeding to the solution of the optimal contract. We assume that the NO considers n aspects of the computing resources provided by FN and will pay the rent based on the QoS of the different aspects.

5.2.1 Operation Cost

In fog computing, the FN encounters both capital expenditure (CapEx) and operational expenditure (OpEx) to provide heterogeneous resources. CapEx is the prominent investment which includes the cost of purchasing and installing equipment such as routers, switches, access points, backhaul aggregators, and the cost of using licensed spectrum issued by the authorities (Cisco 2015a). Meanwhile, the OpEx includes the energy consumption due to signal processing, execution, and data transmission. We assume that the CapEx is fixed, while the OpEx is usage based.

The FN's heterogeneous resources are often measured in disparate scales or units. It has been proved that those measurements can be mapped into one single unit, such as time (Nishio et al. 2013). Thus, by mapping and normalization, we can represent a FN's contribution to resources such as bandwidth, CPU, and transmission power by a vector $a = (a_1, \ldots, a_n)$, $n \geq 1$ for one computing task. After such mapping, each a_i has the same scale or unit and represents one resource type. Such mapping is based on the knowledge that the bandwidth and CPU speed affect the transmission and processing time, respectively. Due to the context aware and location-dependent properties of fog computing, the size of data being processed and the geographic distance between FN and end users also have the impacts on data processing and transmission latency. There are many other aspects that affect the QoS of fog computing that we have not listed in Fig. 5.2, such as transmission power, which can also be mapped to transmission latency.

When providing those resources, the FN's cost incurred is defined in a quadratic form,

$$\psi(a) = \frac{1}{2}a^T C a, \tag{5.1}$$

where C is a symmetric $n \times n$ matrix with the form of

$$C = \begin{bmatrix} c_{11} & \cdots & c_{1n} \\ \vdots & \ddots & \vdots \\ c_{n1} & \cdots & c_{nn} \end{bmatrix}. \tag{5.2}$$

The diagonal element c_{ii} of C reflects the FN's resource-specific cost coefficient, and the off-diagonal elements c_{ij} represent the cost relationship between two resources i and j.

The sign of c_{ij} indicates technologically substitute, complementary, independent between two resources i and j, if $c_{ij} > 0, < 0, = 0$, respectively. If two resources are technologically substitute, raising the quantity of one resource raises the marginal cost of the effort on the other resources. The example of technologically substitute is the relationship between geographic distance and transmission power. To achieve the same data rate at the end user, longer distance requires a higher transmission power consumption. In contrast, raising the quantity on one resource decreases the marginal cost of the other resource if they are technologically complementary. One example is about the relationship between bandwidth and transmission power. Given the same data package size and transmission distance, the larger bandwidth can achieve the same data rate at receiver with lower transmission power. In this example, high quality in one resource eases the cost in the other and thus called technologically complementary. For technologically independent resources, their operation cost is not dependent on the quantity of other resources. There are many technologically independent examples in fog computing, such as the relationships between transmission bandwidth and CPU speed, geographic distance and data size.

In order to lower the mathematical complexity, we only solve the cases without the technologically complementary in this chapter. Thus, the operation cost coefficient matrix is a positive semi-definite matrix with every element in C is nonnegative.

5.2.2 QoS Measurement

The resources such as CPU speed, transmission bandwidth, and power can be easily specified by the FN, and the end user-related parameters such as geographic distance and data size can also be quantified easily, while the measurement of QoS cannot be that accurate. Though those measurements can be mapped onto a timescale that follow the procedure in Nishio et al. (2013), error may come from the design failure of the measurement system.

Given the actual resources provided by FN is a, which is hidden from the NO, the FN's QoS can be observed as a vector of QoS $q = (q_1, \ldots, q_n), n \geq 1$, which can be regarded as the FN's performance in latency reduction. Due to the aforementioned different measurability on QoS, the received information q varies. Therefore, the performance of the FN is a noisy signal of the resources it has provided:

$$q = a + \varepsilon, \tag{5.3}$$

where the random component $\varepsilon = (\varepsilon_1, \ldots, \varepsilon_n), n \geq 1$, is assumed to be normally distributed with mean zero and covariance matrix Σ. Thus, the FN's performance follows the distribution of $q \sim N(a, \Sigma)$.

The variance Σ is a symmetric $n \times n$ covariance matrix with the form of

$$\Sigma = \begin{bmatrix} \sigma_1^2 & \cdots & \sigma_{1n} \\ \vdots & \ddots & \vdots \\ \sigma_{n1} & \cdots & \sigma_n^2 \end{bmatrix}, \tag{5.4}$$

where σ_i^2 denotes the variance of ε_i, and σ_{ij} is the covariance of ε_i and ε_j. The variance denotes the difficulty in guaranteeing the correctness of measuring the QoS and also reflects the resource quantity difference observed at the FN and NO sides. If the variance is large, the measurability of the QoS is difficult, and there is a high probability that the FN's performance is poorly measured and far away from the true amount of resource the FN has provided. If the QoS is easy to measure, the variance will be small or even zero. For example, the data size is an independent measure with variance 0, as well as the geographic distance. While the data processing time depends not only on the data size but also on the complexity of algorithm, which has a large variance, the covariance of two measurements exists because the measurement of one resource may affect the measurement of the others; for example, the transmission time is affected by both bandwidth and power.

5.2.3 Payment Plan

Inspired by the manager's reward package in industry, which comprises a fixed salary, a bonus related to the firm's profits, and stock option-related reward based on the

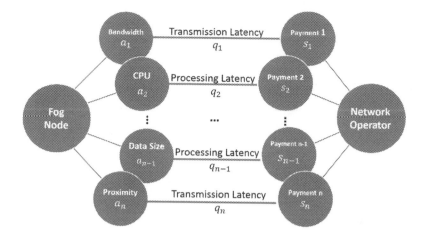

Fig. 5.2 The multi-dimension resource quantity and QoS-consistent payment contract

firm's share price (Bebchuk et al. 2002), we define the FN's payment plan w in fog computing as a linear combination of a fixed salary and QoS-related payments. By restricting the payment plan offered by the NO in the linear form, the payment plan w FN receives by participating in the fog computing can be written as

$$w = t + s^T q, \tag{5.5}$$

where t denotes the fixed salary, which is a constant independent of QoS and regarded as the subscribing fee to complement the FN's CapEx. $s = (s_1, \ldots, s_n), n \geq 1$, is the payment related to the QoS q. As q is a random variable which follows $q \sim N(a, \Sigma)$, the payment plan w is also a random variable with a mean of $t + s^T a$. From the scaling property of covariance, we know that $Var(s^T q) = s^T \Sigma s$. Thus, the payment plan follows the distribution $w \sim N(t + s^T a, s^T \Sigma s)$.

At this point, we can propose the contract that is offered by the NO as (a, t, s), where a and s are $n \times 1$ vectors, and t is a constant value. Under this contract, the NO offers the FN a payment plan, which includes a fixed salary t, and n QoS-related payments $(s_1, \ldots, s_n)$. Figure 5.2 illustrates how this contract works. The FN provides the quantity a_i for resource i in the computing task, which is observed as a QoS q_i by the NO. The NO then offers a payment s_i related to q_i.

5.2.4   Utility of Fog Node

In this model, we assume that the FN has constant absolute risk averse (CARA) risk preferences, which means the FN has a constant attitude toward risk as its income increases. Such a risk preference comes from the FN's concern about its security issue when opening access for end users. Thus, FN utility is represented by a negative exponential utility form (Norstad 1999),

$$u(a, t, s) = -e^{-\eta[w - \psi(a)]}, \tag{5.6}$$

where $\eta > 0$ is the FN's degree of absolute risk aversion

$$\eta = -\frac{u''}{u'}, \tag{5.7}$$

where u is the FN's utility function. A larger value of η means more incentive for the FN to provide more resources for the computing task. The utility and operation cost of the FN are measured in such monetary units that they are consistent with the payment from the NO.

From (5.6), we see that the FN's utility is a strictly increasing and concave function. For lower computation complexity, we can make use of the exponential form of the utility function and use *certainty equivalent* as a monotonic transformation of the FN's expected exponential utility function (Investopedia 2003).

Proposition 5.1 *The FN's utility can be equally represented by certainty equivalent:*

$$CE_u = t + s^T a - \frac{1}{2} a^T C a - \frac{1}{2} \eta s^T \Sigma s. \tag{5.8}$$

From the certainty equivalent, we see the utility consists of the expected payment minus the operation cost and measurement cost.

Proof We have the FN's utility function in (5.6) as $u = -\exp\{-\eta[w - \psi(a)]\}$. From Sect. 5.2, we know that $w \sim N(t + s^T a, s^T \Sigma s)$. As the user incurs an operation cost ψ, the actual income w' has the distribution

$$w' = w - \psi(a) \sim N(t + s^T a - \frac{1}{2} a^T C a, s^T \Sigma s). \tag{5.9}$$

Let μ denotes $t + s^T a - \frac{1}{2} a^T C a$ and σ^2 denotes $s^T \Sigma s$, and we have $w' \sim N(\mu, \sigma^2)$ for simplification. The corresponding density function for w' is

$$f(w') = \frac{1}{\sigma \sqrt{2\pi}} \exp\left[-\frac{(w' - \mu)^2}{2\sigma^2} \right]. \tag{5.10}$$

The corresponding expected exponential utility function is

$$E[u(w')] = -E[\exp(-\eta w')] \tag{5.11}$$

$$= -\int_{-\infty}^{+\infty} \exp(-\eta w') f(w') dw'$$

$$= -\int_{-\infty}^{+\infty} \exp(-\eta w') \frac{1}{\sigma \sqrt{2\pi}} \exp\left[-\frac{(w'-\mu)^2}{2\sigma^2}\right] dw'$$

$$= -\int_{-\infty}^{+\infty} \frac{1}{\sigma \sqrt{2\pi}} \exp\left[-\eta w' - \frac{(w'-\mu)^2}{2\sigma^2}\right] dw'.$$

For the exponential part, we see that

$$-\eta w' - \frac{(w'-\mu)^2}{2\sigma^2} = -\eta w' - \frac{(w'-\mu)^2}{2\sigma^2} + \eta\mu - \eta\mu + \frac{\eta^2\sigma^2}{2} - \frac{\eta^2\sigma^2}{2} \tag{5.12}$$

$$= -\left[\eta w' + \frac{(w'-\mu)^2}{2\sigma^2} - \eta\mu + \frac{\eta^2\sigma^2}{2}\right] - \eta\mu + \frac{\eta^2\sigma^2}{2}$$

$$= -\frac{1}{2}\left[\frac{(w'-\mu)^2}{\sigma^2} + 2\eta(w-\mu) + \eta^2\sigma^2\right] - \eta\mu + \frac{\eta^2\sigma^2}{2}$$

$$= -\frac{1}{2\sigma^2}[(w'-\mu) + \eta\sigma^2]^2 - \eta\mu + \frac{\eta^2\sigma^2}{2}.$$

Thus, the expected exponential utility function becomes

$$E[u(w')] = -\int_{-\infty}^{+\infty} \frac{1}{\sigma \sqrt{2\pi}} \exp\left[-\eta w' - \frac{(w'-\mu)^2}{2\sigma^2}\right] dw' \tag{5.13}$$

$$= -\int_{-\infty}^{+\infty} \frac{1}{\sigma \sqrt{2\pi}} \exp[-\frac{1}{2\sigma^2}[(w'-\mu) + \eta\sigma^2]^2 - \eta\mu + \frac{\eta^2\sigma^2}{2}] dw'$$

$$= -\exp\left(-\eta\mu + \frac{\eta^2\sigma^2}{2}\right) \int_{-\infty}^{+\infty} \frac{1}{\sigma \sqrt{2\pi}} \exp[-\frac{1}{2\sigma^2}[(w'-\mu) + \eta\sigma^2]^2] dw'.$$

As the integration part is the density function of a random variable following a normal distribution with a mean of $\mu - \eta s^2$ and variance σ^2, we have

$$\int_{-\infty}^{+\infty} \frac{1}{\sigma \sqrt{2\pi}} \exp\left[-\frac{1}{2\sigma^2}[(w'-\mu) + \eta\sigma^2]^2\right] dw' = 1 \tag{5.14}$$

Therefore, we have

$$E[u(w')] = \exp\left(-\eta\mu + \frac{\eta^2\sigma^2}{2}\right) = \exp\left[-\eta\left(\mu - \frac{\eta\sigma^2}{2}\right)\right]. \tag{5.15}$$

Therefore, CE represents the same preference as $E[u]$, and the certainty equivalent is a monotonic transformation of the user's expected exponential utility function u. □

5.2.5 Utility of Network Operator

Here, we define the utility of the NO as the expected gross benefits of $V(a)$ minus the payment plan w to the FN. Thus, the NO's expected utility is written as

$$U(a, t, s) = V(a) - w, \tag{5.16}$$

where $V(\cdot)$ is the evaluation function which follows $V(0) = 0$ and $V'(\cdot) > 0$. Different from the FN who has CARA risk preferences, the NO here is assumed to be risk neutral, i.e., $V''(\cdot) = 0$. Thus, the expected profit of the NO can be simplified to

$$U(a, t, s) = \beta^T a - w, \tag{5.17}$$

where $\beta = (\beta_1, \ldots, \beta_n), n \geq 1$, characterizes the marginal effect of the FN's contribution a to the NO's utility $V(a)$. Similar to the definition of FN's certainty equivalent, we can derive the NO's certainty equivalent as

$$CE_p = E[\beta^T a - w], \tag{5.18}$$
$$= \beta^T a - s^T a - t.$$

5.2.6 Social Welfare

With the definitions of both FN's and NO's utility functions and certainty equivalent payoffs, we can have social welfare defined as their joint surplus, i.e., the summation of FN's and NO's equivalent certainty:

$$R = CE_u + CE_p, \tag{5.19}$$
$$= \beta^T a - \frac{1}{2}a^T C a - \frac{1}{2}\eta s^T \Sigma s.$$

The social welfare is the resource provided by the FN minus the operation cost and the cost incurred by inaccurate measurement. Notice that this expression is independent of the fixed salary t, which serves as an intercept term in the contract. Thus, the fixed salary t can only be used to allocate the total certainty equivalent between the two parties (Holmstrom and Milgrom 1991).

5.3   Problem Formulation

With the system model, we can formulate the NO's utility maximization problem while providing the FN necessary incentives to cooperate. The NO's problem can be written as

$$\max_{a,t,s}\quad U(a^*, t, s), \tag{5.20}$$

$$s.t.\quad (a)\ \ a^* \in \arg\max_a u(a, t, s),$$

$$\qquad (b)\ \ u(a^*, t, s) \geq u(\overline{w}),$$

where $u(\overline{w})$ is the reservation utility of the FN when not providing any resource $(a = \mathbf{0})$ in the fog computing. The NO maximizes its own utility under the incentive compatible (IC) constraint (a) that the FN provides the optimal amount of resource a^* maximizing its own utility, and the individual rationality (IR) constraint (b) that the utility FN received is no less than its reservation utility.

Under the assumption of stochastic dependent, the error terms are stochastically interacted, i.e., $\sigma_{ij} \neq 0$. For technologically dependent, we mean that the activities are technologically correlated with each other, i.e., $c_{ij} > 0$ and C is a positive-definite matrix. We solve this multi-dimensional problem by using the certainty equivalent model with the following simple reformulation of the NO's problem:

$$\max_{a,t,s}\quad \beta^T a - s^T a - t, \tag{5.21}$$

$$s.t.\quad (a)\ \ a^* \in \arg\max_a [t + s^T a - \frac{1}{2} a^T C a - \frac{1}{2} \eta s^T \Sigma s],$$

$$\qquad (b)\ \ t + s^T a - \frac{1}{2} a^T C a - \frac{1}{2} \eta s^T \Sigma s \geq \overline{w},$$

where $\overline{w}$ also denotes the reservation utility of the FN when not participating in the fog computing. The IC constraint represents the rationality of the FNs choice of contribution. The IR constraint in (b) ensures that the NO cannot force the FN into accepting the contract.

We first solve the optimal effort by reducing the IC constraint. The FN's certainty equivalent is concave, since its second-order derivative with respect to a is a negative-definite matrix $-C$. Thus, the optimal effort can be determined by taking the first-order derivative of the FN's certainty equivalent regarding a and set $u'(a, t, s) = 0$.

In the matrix differentiation, if we define $\alpha = a^T C a$, as C is a symmetric matrix, we have $\partial \alpha / \partial a = 2a^T C$ (Barzel 1982). Since C is symmetric positive definite, its inverse is existent. Thus, through numerical derivations, we finally have $a = C^{-1}s$ in this multi-dimension case. Accordingly, we substitute the IR constraint in (b) with the optimal amount of resource a^* and simplify the NO's problem to

$$\max_{a,t,s} \quad \beta^T C^{-1} s - s^T C^{-1} s - t, \qquad (5.22)$$

$$s.t. \quad (a) \; t + s^T C^{-1} s - \frac{1}{2}(C^{-1}s)^T C(C^{-1}s) - \frac{1}{2}\eta s^T \Sigma s = \overline{w}.$$

Substituting the value of t in the IR constraint to the objective and differentiating the objective function with respect to s, we have the QoS-related payment s^* in the optimal multi-dimension payment plan as:

$$s^* = (C^{-1} + \eta \Sigma)^{-1} C^{-1} \beta = (I + \eta C \Sigma)^{-1} \beta. \qquad (5.23)$$

With s^*, we have the optimal amount of computing resource in the multi-resource case as

$$a^* = C^{-1}(I + \eta C \Sigma)^{-1} \beta. \qquad (5.24)$$

Representing t by $\overline{w}$, s^*, and a^*, we obtain the fixed payment t in the optimal linear payment plan as:

$$t^* = \overline{w} + \frac{1}{2}s^T (\eta \Sigma - C^{-1})s, \qquad (5.25)$$

$$= \overline{w} + \frac{1}{2}\left[(I + \eta C \Sigma)^{-1}\beta\right]^T (\eta \Sigma - C^{-1}) \left[(I + \eta C \Sigma)^{-1}\beta\right].$$

Using the s^* in (5.23), we can indeed determine how the optimal payment plan varies with the accuracy of QoS measures for each resource and the operation cost coefficient of each resource. For example, when two resources are technologically substitution $c_{ij} > 0$, if the measurability of resource i worsens, that is, σ_i^2 increases, then, as is intuitive, s_j^* goes up, but s_i^* goes down. Thus, there is a measurement complementarity between s_i^* and s_j^* in the presence of technologically substitute problems (Bolton and Dewatripont 2004).

5.4 Simulation Results and Analysis

In this section, we will first give a detailed analysis of how the NO's utility changes by varying the parameters such as the operation cost coefficients and measurement error covariance. Meanwhile, we will conduct a comparison of the NO's utility among different payment plans. To set up the simulation, we assume that the reservation payment of the FN is $\overline{w} = 0$ when not cooperating in the fog computing ($a = 0$). The reason we do not consider the FN's utility is that from the optimal payment plan we have derived, no matter how those parameters change, the FN's utility will remain the same. The optimal payment plan will bring FN the utility the same as the reservation utility $-e^{-\eta \overline{w}}$, which in our case is -1 as we set $\overline{w} = 0$.

In the previous section, we have solved the optimal payment plan when the measurement errors are stochastic dependent and the resource types are technologically dependent. As this multi-dimension case is the most general case in reality, we name this mechanism by *General*. For comparison, we propose 5 more payment plans. The first one is the optimal payment plan when the measurement error and resource type are independent, and thus, we name it by *Independent*. The second payment plan is called *Single Bonus* that is the payment plan obtained in the one-dimension case. In this one-dimension case, we can regard the NO payments FN on the QoS of a single one resource type, for example, measuring the data processing latency by only taking the CPU speed into account. The third and forth ones are special cases of the *General*: One is stochastic independent but technologically dependent, and the other one is technologically independent but stochastic dependent, which are named by *Stochastic Independent* and *Technologically Independent*, respectively. The last one is called *Opening Reward*, which is the payment plan only contains a fixed salary t. We can regard this mechanism as the NO offers the FN a one-time payment at subscription. But this *Opening Reward* mechanism does not care about FN's future service quality.

In Fig. 5.3, we compare the NO's utility from the six payment plans as we vary the resource-specific operation cost coefficient c_{ii}. From the simulation results, we see

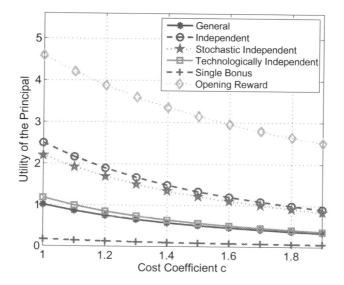

Fig. 5.3 The NO's utility as cost coefficient c_{ii} varies

that as the cost coefficient c_{ii} increases, the NO's utility is decreasing in contrast. The reason for this phenomenon is that a larger cost coefficient c_{ii} means more operation cost when providing such a resource. Therefore, the FN is less likely to contribute in the fog computing. With less computing resources provided by the FN, the QoS will decrease and the NO's utility will certainly decrease. In addition, from Fig. 5.3, we see that the NO obtains the largest utility in the *Opening Reward* case. Followed by the *Independent, Stochastic Independent*, and *Technologically Independent*, the *General* case proposed by us brings the fifth highest utility to the NO, while the *Single Bonus* gives the least utility.

In Fig. 5.4, we analyze the impact of FN's risk averse degree η on the NO's utility. As the NO's utility $V = a - t$ in the *Opening Reward* is independent of the risk averse degree η, we cannot see any change in the NO's utility. For the other five payment plans, we see that the NO's utility is decreasing as the FN's risk averse degree η increases. This result is intuitive as a larger η means the FN becomes more conservative and sensitive to risk, thus less likely to open access to end users. With fewer resources obtained from the FN, the NO's utility will certainly decrease. From Fig. 5.4, we also obtain the similar ranking of the NO's utility as in the previous figure: The *Independent* case brings higher utility than the *Stochastic Independent, Technologically Independent*, and *General* one, and the *Single Bonus* one brings the smallest utility for the NO.

In Fig. 5.5, we increase the variance σ_i^2 to see how the NO's utility varies. Similar to the previous case, the NO's utility $V = a - t$ in the *Opening Reward* is independent of the covariance matrix. Thus, we cannot see any change in the NO's utility. For the other payment plans, the NO's utility is decreasing in the variance, which

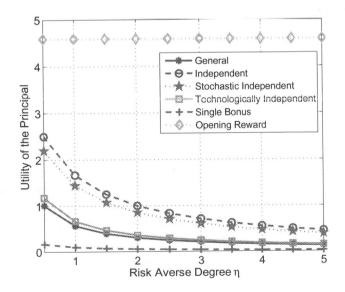

Fig. 5.4 The NO's utility as risk averse degree η varies

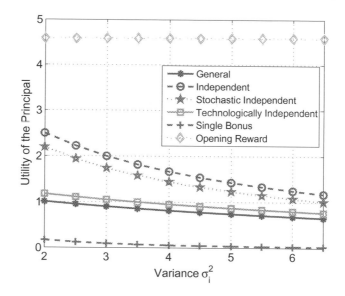

Fig. 5.5 The NO's utility as measurement error variance σ_i^2 varies

is in accordance with our conclusion in the previous section. The variance σ_i^2 of measurement error denotes the relationship between the resources provided by the FN and the QoS observed by the NO. As σ_i^2 increases, it indicates a weaker relationship between resource quantity and the expected QoS achieved. As a result, the FN is likely to provide less amount of resource with increases in uncertainty and thus a lower cost of cooperation. With the decrease in computing resources, the QoS is lowered and the NO's utility will certainly decrease. From Fig. 5.5, we also obtain the similar ranking of the NO's utility as in the previous figure: The *Independent* case brings higher utility than *Stochastic Independent*, followed by *Technologically Independent* and *General* one, and the *Single Bonus* one brings the lowest utility for the NO.

The reason for the quality ranking of the six payment plans shown in Figs. 5.3, 5.4, and 5.5 is as follows. The *Independent* payment plan is the ideal case of the *General* multi-dimension case. As less measurement cost is occurred when predicting the QoS and less operation cost is encountered due to technology substitution, a higher utility is obtained than the other payment plans. The *Stochastic Independent* and *Technologically Independent* are partial independent cases of the *General* multi-dimension one; thus, the NO's utility lies between the *Independent* and *General*. But as we have assigned larger values for the covariance matrix of the measurement error than the operation cost coefficient matrix, more computing resources will be provided in the *Stochastic Independent* than in the *Technologically Independent*. Therefore, the NO's utility is higher in the Stochastic Independent than in the Technologically Independent case. While the Single Bonus provides the FN with payment that evaluates only one type of resource, thus less incentive is induced and lower utility is received. As a result, the FNs have less incentive to provide computing resources. In return, less utility is obtained by the NO. For the result of the *Opening Reward* case, it seems unreasonable at the first sight, as it brings the NO the highest utility than the other payment plans. While we notice that *Opening Payment* is a "once-for-all" deal which does not provide continuous incentives for the FNs, i.e., after the FN has finished the computing task and received the payment, it is likely to stop cooperating in the future.

5.5 Summary

In this chapter, we have investigated the problem of maximizing NO's revenue by efficiently allocating FNs' computation resources in fog computing. The optimal payment plan is solved by paying the rent of FN's computing resources from a multi-dimension evaluation while ensuring the FN's cooperation. Furthermore, we use the numerical results to analyze the optimal payment plan by varying different parameters. In addition, we compare the NOs' utility under the six different payment plans and show that the NO's utility deteriorates into large operation cost coefficient, higher risk aversion of FNs, and large measurement error variance. By this end, we have provided applications with *adverse selection* only or *moral hazard* only in wireless networks. In the next chapter, we will consider one application in cognitive radio network that has both *adverse selection* and *moral hazard* problems.

References

Barbarossa S, Sardellitti S, Lorenzo PD (2014) Fog computing: will it be the future of cloud computing? IEEE Signal Process Mag 31(6):45–55

Barzel Y (1982) Measurement cost and the organization of markets. J Law Econ 25(1):27–48

Bebchuk LA, Fried JM, Walker D (2002) Managerial power and rent extraction in the design of executive compensation. Univ Chicago Law Rev 69(2):751–846

Bolton P, Dewatripont M (2004) Contract theory. The MIT Press, Cambridge, MA

Bonomi F, Milito R, Zhu J, Addepalli S (2012) Fog computing and its role in the internet of things. In: Proceedings of the first edition of the MCC workshop on mobile cloud computing, pp 13–16

Cisco (2015a) Cisco fog computing solutions: unleash the power of the internet of things. Technical report

Cisco (2015b) Fog computing and the internet of things: extend the cloud to where the things are. Technical report

Holmstrom B, Milgrom P (1991) Multitask principal-agent analyses: incentive contracts, asset ownership, and job design. J Law Econ Organ Sp(3):24–52

Investopedia (2003) Certainty equivalent. http://www.investopedia.com/terms/c/certaintyequivalent.asp

Nishio T, Shinkuma R, Takahashi T, Mandayam NB (2013) Service-oriented heterogeneous resource sharing for optimizing service latency in mobile cloud. In: Proceedings of the first international workshop on mobile cloud computing and networking, Mobilecloud '13, pp 19–26

Norstad J (1999) An introduction to utility theory. Technical report. http://www.norstad.org/finance/util.pdf

Patel M, Hu Y, Hédé P, Joubert J, Thornton C, Naughton B, Ramos JR, Chan C, Young V, Tan SJ, Lynch D, Sprecher N, Musiol T, Manzanares C, Rauschenbach U, Abeta S, Chen L, Shimizu K, Neal A, Cosimini P, Pollard A, Klas G (2014) Mobile-edge computing. Introductory technical white paper

Sardellitti S, Scutari G, Barbarossa S (2015) Joint optimization of radio and computational resources for multicell mobile-edge computing. IEEE Trans Signal Inf Process Netw 1(2):89–103

Werin L, Wijkander H (1992) Contract economics. Blackwell Publishers, Oxford, UK

Chapter 6
Financing Contract with Adverse Selection and Moral Hazard for Spectrum Trading in Cognitive Radio Networks

6.1 Introduction

After discussing the applications of the two basic problems, *adverse selection* and *moral hazard*, we now proceed to the mixed problem in wireless networks when both are present. The recent popularity of handheld mobile devices, such as smartphones, enables the interconnectivity among mobile users without the support of Internet infrastructure. With the wide use of such applications, the data outburst leads to a booming growth of various wireless networks and a dramatic increase in demand for radio spectrum (Letaief and Zhang 2009). However, we are currently in the exhaustion of the available spectrum. Thus, cognitive radio (CR) has emerged as a new design paradigm as its opportunistic access to the vacant licensed frequency bands, which releases the spectrum from shackles of authorized licenses and, at the same time, improves the spectrum utilization efficiency (Kim and Shin 2008).

Cognitive radio networks (CRNs) are designed based on the concept of dynamic spectrum sharing where CR users can opportunistically access the licensed spectrum (Hossain et al. 2009). In a CRN, the primary users (PUs) are licensed users to utilize the frequency band, the secondary users (SUs) can only utilize those spectrum resources when the PUs are vacant. Whenever the PUs are back, the SUs must vacate the frequency band immediately to guarantee the PUs' quality of service (QoS) (Zhu and Yum 2007). In other words, in a CRN, the PUs have higher priority to use the frequency bands than the SUs. The SU can be regarded as a radio which is capable of changing its transmitter parameters and transmitting/operating frequency based on its interaction with the environment (Brodersen et al. 2004).

In CRNs, the problems of spectrum sensing and resource allocation have been extensively studied in previous works such as Wang et al. (2010a). In this work, we will focus on the economic aspect of spectrum trading between PU and SU, which achieves SU's dynamic spectrum accessing/sharing and creates more economically benefits for the PU. The idea of the market-driven structure has initiated the spectrum trading model in CRNs and promoted a lot of interesting researches on the design of trading mechanisms. Through spectrum trading, PUs can sell/lease their vacant

© Springer International Publishing AG 2017
Y. Zhang and Z. Han, *Contract Theory for Wireless Networks*,
Wireless Networks, DOI 10.1007/978-3-319-53288-2_6

spectrum for monetary gains, and SUs can purchase/rent the available licensed spectrum if they are in need of radio resources to support their traffic demands (Gao et al. 2011).

However, most mechanisms such as Pan et al. (2012) are designed for the one-shot trading problem. Different from the previous studies, we consider offering a contract-based mechanism that allows the SU to do a financing, as we do for a house or a car (Laffont and Tirole 1988). That is, the SU only needs to pay part of the total amount at the point of signing the contract, known as the down payment. Then, the spectrum can be released to the SU by the PU. Successively, the SU can utilize the spectrum to transmit package and generate revenue. Afterward, the SU pays the rest of the loan, known as the installment payment.

To obtain the optimal contract, the PU must consider the SU's current and future financial status (Scott 2014). When the SU utilizes the spectrum to generate revenue, however, the PU may not have the knowledge of the SU's capability, i.e., what is the SU's probability of successful making profit, in which case the problem of *adverse selection* arises (Akerlof 1995). Moreover, the PU neither knows how much effort the SU exerts, where the problem of *moral hazard* arises (Roland 2000). Thus, we model the spectrum trading by a contract theoretical model which involves both *adverse selection* and *moral hazard* problems as shown in Fig. 6.1.

The main contributions of this chapter are as follows.

- A financing contract for spectrum trading is proposed, instead of a one-shot trading.
- The innovative model of the financing contract that involves both *adverse selection* and *moral hazard* problems is considered.
- The solutions to the problems under three different scenarios, i.e., the general case where both *adverse selection* and *moral hazard* are present, the two extreme cases where only *adverse selection* or *moral hazard* is present.
- The analysis of how *adverse selection* and *moral hazard* affect the SU's activity and PU's contract design is provided.
- Numerical results that are provided to compare the optimal contracts under the three scenarios and to study the key parameters' influences on the PU's and SU's payoffs.

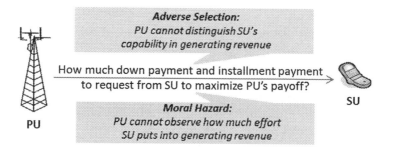

Fig. 6.1 The problems of *adverse selection* and *moral hazard* in financing contract design

The remainder of this chapter is organized as follows. First, we will introduce the system model in Sect. 6.3. Then, a literature review of spectrum trading and contract theory application in wireless networks is conducted in Sect. 6.2. Next, the system model is described in Sect. 6.3, and we formulated the PU's payoff maximization problems under the three scenarios in Sect. 6.4. The performance evaluation is conducted in Sect. 6.5. Finally, Sect. 6.6 summarizes the chapter.

6.2 Related Works

Spectrum trading in CRNs has been extensively studied by using game theory (Han et al. 2011). Different game theoretical models have been adopted, such as potential game (Nie and Comaniciu 2005), evolutionary game, non-cooperative game (Niyato et al. 2009), and Stackelberg game (Xie et al. 2012). Despite game theory, auction theory is another popular method to solve the spectrum trading problem. Despite game theory, auction theory is another popular method to solve the spectrum trading problem, such as double auction adopted in Zhou and Zheng (2009). The fundamental one-shot auction has been extended to real-time fast auction algorithms by Gandhi et al. (2007) and performance-related auction by Wang et al. (2010b). Although one-shot auction-based spectrum trading has been extensively studied, there is a few works that has tangled the non-cash auction.

Contract theory has recently emerged into spectrum trading by some peer works. As far as we know, contract theory is first used to solve the problem of spectrum sharing in cognitive radio network (CRN) by Gao et al. (2011). In this work, a primary user (PU) acts as a seller who sets the spectrum trading contract as *(qualities, prices)*, and the second users (SUs) act as a buyer to choose a contract to sign. Another application in CRNs can be found in Duan et al. (2014), in which the authors model the PU and SUs as employer and employees, respectively. Then designing the *(performance, reward)* in contract as *(relaying power, spectrum accessing time)*, so that SUs will be rewarded with certain spectrum assessing time if they satisfied the relaying power requirement of the PU.

Despite the previous two works that applied contract theory into spectrum trading, some other areas have also been explored. Duan et al. (2012) designed incentive mechanisms for smartphone users' collaboration on both in data acquisition and distributed computing. The SP acts as an employer and smartphone users will be employees. Rewards will be paid according to the amount of data collected and distributed computing users made. In the OFDM-based cooperative communication system, (Hasan and Bhargava 2013) used contract theory to tackle the source node's relay selection problem. The offers/contracts consist of a menu of desired signal-to-noise ratios (SNRs) at the destination and corresponding payments. In our previous work (Zhang et al. 2015b), we applied the *adverse selection* model in cellular traffic off-loading through D2D communication, by offering rewards to encourage content owners to participate and cooperate with other devices via D2D. We modeled the BS as employer and D2D user as employee and solved contract bundle with a required performance and an absolute performance-related reward. The performance is defined as a certain data rate that the UE must provide during the D2D communication.

However, all of the above works fall into the applications of *adverse selection* problems in wireless networks. Compared to the wide adoption of the *adverse selection* problem, the *moral hazard* problem has hardly been applied in wireless networks by now. However, having seen a great potential of this model, we have done some preliminary applications in mobile crowdsourcing. As mentioned in the beginning of this survey, many users hesitate to participate in mobile crowdsourcing with certain concerns, which results in serious impediment to the exploitation of location-based services. By adopting the *moral hazard*, the incentive mechanism can be designed by regarding the SP "employs" a large group of users to upload location-based data and reward them by their performance. Thus, with the large group of users as employees, the multilateral *moral hazard* model can be applied. In another work of ours (Zhang et al. 2015a), we consider the mobile users competing in the crowdsourcing to win reward as in a tournament, and they are rewarded by their rank orders, i.e., relative performance.

As we see, the literature in contract theory applied wireless networks, either *adverse selection* or *moral hazard* is considered when modeling the problem. In practice, however, it is usually hard to decide which of the two problems is more important, i.e., to figure out if it is a *moral hazard* problem or an *adverse selection* problem (Edward and Prescott 1984). Indeed, most incentive problems in practice are the combinations of *moral hazard* and *adverse selection*.

6.3 System Model

Based on the earliest model in Laffont and Tirole (1988), we consider the spectrum trading between one PU and one SU in one CRN. The contract can be extended to other SUs in the same CRN. Both the PU and SU are risk-neutral which means they have no preference between saving and consuming. The PU's spectrum is vacant, and the PU cannot generate any revenue from the vacant spectrum unless selling/leasing to the SU.

The PU offers a financing plan (t_i, r_i) to the SU to pay for utilizing the spectrum, where t_i is a down payment and r_i is an installment payment to be paid from future revenues generated. The problem that the PU needs to solve is to find the optimal contract that can maximize its expected return from the spectrum trading by deciding how much down payment, and how much installment payment the SU needs to pay.

The SU makes use of the spectrum to run its own "business," which can only result in a success (receive a revenue of $R \geq r_i \geq 0$) or failure (receive a revenue of 0), i.e., the revenue realizations: $X \in \{0, R\}$. The SU may be more or less able at utilizing this vacant spectrum, whose capability may belong to two different types $\theta \in \{\theta_L, \theta_H\}$ with $\theta_L < \theta_H$, which donate lower or higher capability to generate revenue, respectively. The PU may not be able to observe the SU's capability type, but with a priori that the SU has a high-capability θ_H with probability $\beta \in [0, 1]$ and a low-capability θ_L with probability $(1 - \beta)$.

The SU's capability θ can be translated into the probability of getting the high revenue R. Besides the capability, the SU can also increase its efforts e (e.g., transmission power) to raise the probability of getting R. Thus, we define the SU's probability of generating high revenues R as $\theta e \in (0, 1)$. In addition, we assume that the SU's operation cost ψ on the spectrum is a convex function of effort e, which is

$$\psi(e) \equiv \frac{c}{2}e^2, \tag{6.1}$$

where c is the cost coefficient. To ensure that the probability $0 < \theta e < 1$, we take c to be large enough that the SU would never want to choose a level of effort e such that $\theta e \geq 1$.

We assume that there is no installment payment if the SU cannot generate revenue from utilizing the spectrum, i.e., $r_i = 0$ if $X = 0$. The installment payment r_i is made only when $X = R$. Thus, the expected payoff of SU with capability θ_i under contract (t_i, r_i) then takes the form of

$$U_{SU_i} = \theta_i e_i(R - r_i) - t_i - \frac{c}{2}e_i^2, \quad i \in \{L, H\}. \tag{6.2}$$

The revenue R minus installment payment r_i is the SU's income. The SU's expected payoff is the expected income minus the down payment and cost of operation.

Similarly, we define the expected payoff of the PU as

$$U_{PU} = \sum_i \beta_i(t_i + \theta_i e_i r_i), \quad i \in \{L, H\}, \tag{6.3}$$

$$= \beta[t_H + \theta_H e_H r_H] + (1 - \beta)[t_L + \theta_L e_L r_L].$$

The PU's expected payoff is the summation of the down payment and expected installment payment.

6.4 Problem Formulation

In this section, we will solve the PU's problem by considering three scenarios, i.e., the general case where both *moral hazard* and *adverse selection* are present, the two extreme cases where only *moral hazard* or *adverse selection* is present, respectively

6.4.1 PU's Payoff Maximization Problem

The PU's problem payoff maximization problem is

$$\max_{(t_i,r_i)} \beta[t_H + \theta_H e_H r_H] + (1 - \beta)[t_L + \theta_L e_L r_L], \tag{6.4}$$

$$s.t.$$

$$(IC) \quad \theta_i e_i (R - r_i) - t_i - \frac{c}{2}e_i^2 \geq \theta_i e_i'(R - r_j) - t_j - \frac{c}{2}e_i'^2,$$

$$(IR) \quad \theta_i e_i (R - r_i) - t_i - \frac{c}{2}e_i^2 \geq 0,$$

$$\forall j \neq i, \quad i, j \in \{L, H\},$$

where e_i' is the effort of θ_i SU when selecting contract (t_j, r_j). The IC constraint stands for incentive compatibility, which means the SU can only maximize its expected payoff by selecting the financing contract that fits its own capability. The IR constraint stands for individual rationality, which provides the SU necessary incentives to sign the contract.

Taking the first derivative of SU's expected payoff with respect to effort e, we have the SU's optimal choice of effort e^* under the contract (t, r) as

$$e_i^* = \frac{1}{c}\theta_i(R - r_i), \quad i \in \{L, H\}. \tag{6.5}$$

Similarly, we have $e_i'* = \frac{1}{c}\theta_i(R - r_j)$. As we can see from this equation, the SU's optimal choice of effort e_i^* is independent of t_i but is decreasing in r_i. In other words, the SU will have fewer incentives to put more effort in utilizing the spectrum, if it must share more of the generated revenue, regardless of the amount of the down payment t_i. The decrease of effort e directly affects the probability of successfully generating revenue R. Thus, it is critical to balance the trade-off between providing necessary incentives for the SU and request more installment payment from the SU.

Replacing SU's choice of effort e_i and e_i' in (6.4), we have the PU's problem in the following form.

$$\max_{(t_i,r_i)} \beta[t_H + \frac{1}{c}\theta_H^2(R - r_H)r_H] + (1 - \beta)[t_L + \frac{1}{c}\theta_L^2(R - r_L)r_L], \tag{6.6}$$

$$s.t. \quad (IC) \quad \frac{1}{2c}[\theta_i(R - r_i)]^2 - t_i \geq \frac{1}{2c}[\theta_i(R - r_j)]^2 - t_j,$$

$$(IR) \quad \frac{1}{2c}[\theta_i(R - r_i)]^2 - t_i \geq 0,$$

$$\forall j \neq i, \quad i, j \in \{L, H\}.$$

In this problem, it is not possible to decide on a priority which of the two incentive problems is the more important, i.e., to disentangle the *moral hazard* from the *adverse selection* dimension. In the following section, we will detail the respective roles of *moral hazard* and *adverse selection* and the implications of their simultaneous presence. As we shall see, the design of the optimal financing contract for this problem depends on whether only the *adverse selection* or the *moral hazard* (or both) is explicitly taken into account.

6.4.2 Optimal Contract with Moral Hazard only

Suppose that the PU is able to observe the SU's financial status, so that the *adverse selection* problem is removed, and the only remaining incentive problem is *moral hazard*. Then, the PU's problem can be treated separately for different capability SU and reduces to

$$\max_{(t_i, r_i)} t_i + \frac{1}{c}\theta_i^2 (R - r_i)r_i, \tag{6.7}$$

$$s.t. \ (IR) \ \frac{1}{2c}[\theta_i(R - r_i)]^2 - t_i \geq 0,$$

$$i \in \{L, H\}.$$

Since the IR constraint is binding, the problem becomes

$$\max_{r_i} \frac{1}{2c}[\theta_i(R - r_i)]^2 + \frac{1}{c}\theta_i^2(R - r_i)r_i. \tag{6.8}$$

After simplification, the problem is equivalent to

$$\max_{r_i} \frac{1}{2c}\theta_i^2(R^2 - r_i^2). \tag{6.9}$$

The solution for this maximization problem is $r_H = r_L = 0$ and $t_i = \frac{1}{2c}\theta_i^2 R^2$. As there is no *adverse selection* present, the PU only needs to minimize the negative effect of *moral hazard*. To avoid the *moral hazard* problem, it is optimal for the PU to sell the spectrum for cash only and not keep any financing participation in. The only reason why the PU might want to keep some financing participation in this pure *moral hazard* case is that the SU may be financially constrained and may not have all the cash available for the down payment.

6.4.3 Optimal Contract with Adverse Selection only

Suppose now that the SU's effort level is fixed at some level $\widehat{e}$, but the PU cannot observe the SU's financial status. The PU's problem is then reduced to

$$\max_{(t_i,r_i)} \beta[t_H + \theta_H \widehat{er}_H] + (1 - \beta)[t_L + \theta_L \widehat{er}_L], \qquad (6.10)$$

$$s.t. \quad (IC) \quad \theta_i \widehat{e}(R - r_i) - t_i \geq \theta_i \widehat{e}(R - r_j) - t_j,$$

$$(IR) \quad \theta_i \widehat{e}(R - r_i) - t_i - \frac{c}{2}\widehat{e}^2 \geq 0,$$

$$\forall j \neq i, \quad i, j \in \{L, H\}.$$

This problem also has a simple solution: $r_H = r_L = R$ and $t_i = -\frac{1}{2}c\widehat{e}^2 < 0$. Intuitively, the down payment should be larger than or equal to 0. However, in this optimal contract, the SU has a negative down payment, i.e., the PU has to pay $\frac{1}{2}c\widehat{e}^2$ to the SU instead. This result is due to the fact that the PU asks for 100% of the future return from the SU. In order to hold the IR constraint, a down payment from the PU to the SU is necessary.

6.4.4 Optimal Contract with both Adverse Selection and Moral Hazard

The simplicity of the preceding solutions is of course driven by the extreme nature of the setup. However, neither extreme formulation is an adequate representation of the basic problem in practice and that it is necessary to allow for both types of incentive problems to have a plausible description of the spectrum trading in practice. Not surprisingly, the optimal menu of contracts where both types of incentive problems are present is some combination of the two extreme solutions that we have highlighted.

Solving the problem of the PU can be done by relying on the pure *adverse selection* methodology detailed in Bolton and Dewatripont (2004). Specifically, the analysis shows that only the IR constraint of the θ_L SU and the IC constraint of the θ_H SU will be binding. Indeed, first, when the θ_L SU earns nonnegative rents, so will the θ_H SU, who can always mimic the θ_L SU. Second in the symmetric information scenario, that is, pure *moral hazard* optimum, the PU manages to leave the θ_H SU with no rents, but this outcome is what would induce the SU to mimic the θ_L SU. Therefore, the PU has to solve

$$\max_{(t_i,r_i)} \{\beta[t_H + \frac{1}{c}\theta_H^2(R - r_H)r_H] + (1 - \beta)[t_L + \frac{1}{c}\theta_L^2(R - r_L)r_L]\}, \qquad (6.11)$$

$$s.t.$$

$$(IC) \quad \frac{1}{2c}[\theta_H(R - r_H)]^2 - t_H = \frac{1}{2c}[\theta_H(R - r_L)]^2 - t_L,$$

$$(IR) \quad \frac{1}{2c}[\theta_L(R - r_L)]^2 - t_L = 0.$$

Using the two binding constraints to eliminate t_H and t_L from the objective function, we obtain the usual efficiency-at-the-top condition $r_H = 0$ (as in the pure *moral hazard* case).

The first-order condition with respect to r_L involves the usual trade-off between surplus extraction from the θ_L SU and informational rent concession to the θ_H SU and leads to

$$r_L = \frac{\beta(\theta_H^2 - \theta_L^2)R}{\beta(\theta_H^2 - \theta_L^2) + (1 - \beta)\theta_L^2}, \tag{6.12}$$

which is bigger than 0.

By taking r_L and r_H into the constraints IC and IR in (11), we obtain the down payments in this general case, which are

$$t_L = \frac{1}{2c}[\theta_L(R - r_L)]^2, \tag{6.13}$$

$$t_H = t_L + \frac{1}{2c}\theta_H^2[(R - r_H)^2 - (R - r_L)^2]. \tag{6.14}$$

The optimal menu of contracts is thus such that there is no effort-supply distortion for the high-capability SU because it is a 100% residual claimant. But there is a downward effort distortion for the low-capability SU that serves the purpose of reducing the informational rent of the high-capability SU. The extent of the distortion, measured by the size of r_L, depends on the size of the ability differential $(\theta_H^2 - \theta_L^2)$ and on the PU's prior β: The more confident the PU is that it faces a high-SU type, the larger is its stake r_L and the larger is the down payment t_H.

6.5 Simulation Results and Analysis

In this section, we will first give an analysis about the financing contract when both *adverse selection* and *moral hazard* are present by varying the parameters such as the cost coefficient, revenue, and the SU's probability of being θ_H. For the two extreme cases where only *adverse selection* or *moral hazard* is present, the results can be predicted from the general case. Then, we will conduct comparisons among the PU's and SU's payoffs, and social welfare among the three scenarios we have proposed. In the simulation setup, we assume that $\theta_H = 2$ and $\theta_L = 1$. We set $c = 10$ as a high value so that we can guarantee $\theta_H e < 1$ always holds.

6.5.1 Financing Contract Analysis

In Fig. 6.2, we show the financing contract for θ_H SU when both *adverse selection* and *moral hazard* are present. We see that, with the varying of the three parameters,

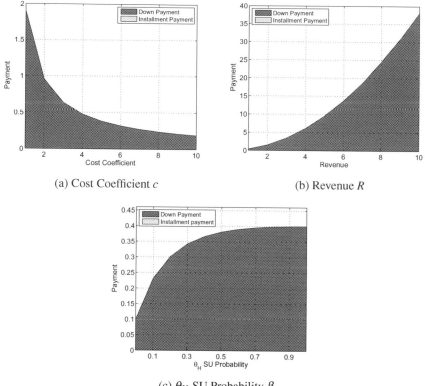

(a) Cost Coefficient c (b) Revenue R

(c) θ_H SU Probability β

Fig. 6.2 The financing contract for θ_H SU as parameters vary

the installment payment r_H remains 0, as we have stated in the previous section. When the PU knows it is facing a SU with enough cash in hand, it will ask the SU to pay the total amount money when signing the contract, but no installment payment afterward.

From Fig. 6.2a, we see that as the cost coefficient c increases, the down payment (i.e., the price of the spectrum) decreases. This result is intuitive in the sense that when the SU's cost of generating revenue by utilizing the spectrum increases, the SU will be less likely to participate. Thus, the PU must lower its price to attract SU's participation. Otherwise, the vacant spectrum is wasted and 0 payoff is obtained by the PU.

In Fig. 6.2b, we see that as the SU's revenue R by "running" on the PU's spectrum increases, the cash payment required from the PU increases. This result is also easy to see as if the spectrum can bring more revenue for the SU, the spectrum's value is higher. Thus, the PU would definitely assign a higher price for the spectrum.

Figure 6.2c shows when the PU's probability of trading with a θ_H SU increases, it will also raise the spectrum's price. As we have defined in the system model, the

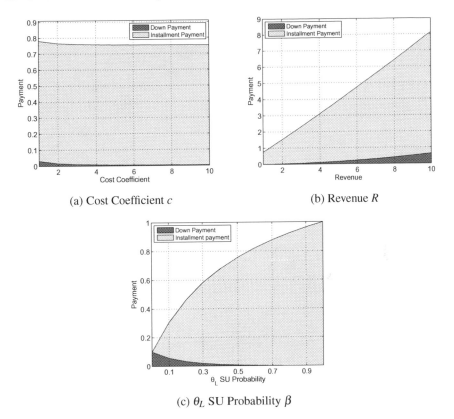

(a) Cost Coefficient c

(b) Revenue R

(c) θ_L SU Probability β

Fig. 6.3 The financing contract for θ_L SU as parameters vary

SU's successful probability of obtaining a revenue is θe. Therefore, under the same effort e, the highly capable SU will bring larger expected revenue than low-capability SU, as $\theta_H > \theta_L$. Thus, similar to Fig. 6.2b, the PU will raise the price as the value of spectrum increases.

Figure 6.3 is similar to Fig. 6.2, as we are showing the financing contract for the θ_L SU, while different from Fig. 6.2 is that the PU asks for both cash and installment payment from the low-capability SU, instead of only down payment when the SU is high capable. This result is intuitive in the sense that the low-capability SU has limited cash at hand for the trading. Thus, the PU will only ask for a small amount of down payment first, while most of the money is paid after the SU has generated revenue from using the spectrum as we have stated in the previous section.

From Fig. 6.3a, we see that as the cost coefficient c increases, both the down and installment payments decrease. The reason for this result is the same as shown in Fig. 6.2a that the PU must lower its price to attract SU's participation.

In Fig. 6.3b, we see that as the SU's revenue R by running on the PU's spectrum increases, both the down and installment payment asked from the PU increase. The reason for this result is the same as shown in Fig. 6.2b that as the spectrum's value grows higher, the PU would definitely ask for a higher price.

Figure 6.3c shows the optimal contract when the PU's probability of trading with a θ_L SU increases. As the PU becomes more certain that it is trading with a low-capability SU with less cash in hand, it will lower the cash payment first, but ask for more installment payment instead, which is the SU's price of paying less cash at first.

6.5.2 System Performance

From Figs. 6.4, 6.5, and 6.6, we compare the system performance under the three scenarios we have proposed: *moral hazard* only, *adverse selection* only, and when both are present. In the following part, we will give a detailed analysis of the cost coefficient c, revenue R, and distribution β's effects on the system performance.

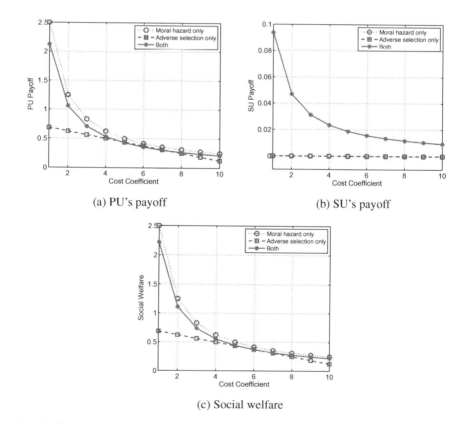

(a) PU's payoff   (b) SU's payoff

(c) Social welfare

Fig. 6.4 The system performance as the cost coefficient c varies

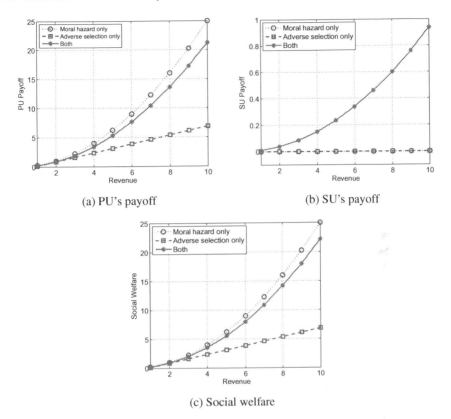

(a) PU's payoff   (b) SU's payoff

(c) Social welfare

Fig. 6.5 The system performance as the revenue R varies

6.5.2.1   Cost Coefficient

In Fig. 6.4, we vary the value of the cost coefficient c to see the effects on the PU's and SU's payoffs, and the social welfare of the three scenarios. We can see that PU's and SU's payoffs and social welfare decrease as the cost coefficient increases, except the SU's payoff under *moral hazard* only and *adverse selection* only scenarios. Under those two extreme cases, the PU has the full acknowledgment of either the SU's cash in hand or the effort put into using the spectrum. Thus, the PU can extract as much revenue as possible from the SU, which leaves the SU with 0 payoff. The reason for the decreasing of payoffs and social welfare is similar to the analysis we gave for Figs. 6.2a and 6.3a that as the cost increasing, the price for the spectrum will decrease to attract SU. As a result, the payoffs of the PU and SU, together with social welfare, will decrease.

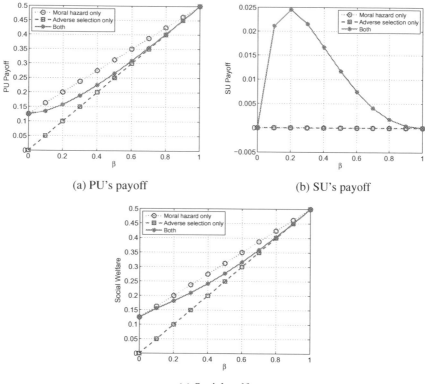

(a) PU's payoff   (b) SU's payoff

(c) Social welfare

Fig. 6.6 The system performance as the θ_H SU probability β varies

6.5.2.2 Revenue

In Fig. 6.5, we try to see the PU and SU's payoffs, and the social welfare when the revenue R can be generated from using the spectrum increases. We see that the payoffs and social welfare increase with the revenue except the SU's payoff under *moral hazard* only and *adverse selection* only scenarios. The increase of payoffs and social welfare with the revenue R is easy to understand as we have explained in the previous paragraph that the PU will extract all the information rent from the SU.

6.5.2.3 Distribution

In Fig. 6.6, we see that PU's payoff and the social welfare increase as β gets larger. The reason for this result is the same as we have explained for Figs. 6.2c and 6.3c, as the PU will ask for more money if it believes that it is facing a high-capability

SU. However, the increase of β has a negative effect on the SU's payoff as the PU is trying to extract revenue from the SU.

Overall, from Figs. 6.4, 6.5, and 6.6, we see that the two extreme cases serve as the upper and lower bounds, respectively. The PU's payoff in the general case where both *moral hazard* and *adverse selection* present lies between the two extreme cases.

6.6   Summary

In this chapter, we have proposed a financing contract to address the problem of spectrum trading in a cognitive radio network. We have modeled the problem by considering both the *adverse selection* and *moral hazard* of the SU. In addition, we have analyzed three scenarios, i.e., two extreme cases where only *adverse selection* or *moral hazard* is present, and the general case where both are present. Through simulations, we have shown different parameters' effects on the system performance, and that the two extreme cases serve as the upper and lower bounds for the general case with both problems present. By this end, we have provided five applications of complete contracts in wireless networks. In the next chapter, we will move forward to the incomplete contract application in mobile, virtualized network.

References

Akerlof G (1995) Essential readings in economics. Macmillan Education UK, London. Chap The market for "Lemons": quality uncertainty and the market mechanism, pp 175–188

Bolton P, Dewatripont M (2004) Contract theory. The MIT Press, Cambridge, MA

Brodersen RW, Wolisz A, Cabric D, Mishra SM (2004) Corvus: a cognitive radio approach for usage of virtual unlicensed spectrum. white paper. http://bwrc.eecs.berkeley.edu/Research/MCMA?destination=Research

Duan L, Kubo T, Sugiyama K, Huang J, Hasegawa T, Walrand J (2012) Incentive mechanisms for smartphone collaboration in data acquisition and distributed computing. In: Proceedings of IEEE INFOCOM, Orlando, FL

Duan L, Gao L, Huang J (2014) Cooperative spectrum sharing: a contract-based approach. IEEE Trans Mobile Comput 13(1):174–187

Edward C, Prescott RMT (1984) Pareto optima and competitive equilibria with adverse selection and moral hazard. Econometrica 52(1):21–45

Gandhi S, Buragohain C, Cao L, Zheng H, Suri S (2007) A general framework for wireless spectrum auctions. In: 2nd IEEE international symposium on new frontiers in dynamic spectrum access networks (DySPAN), pp 22–33

Gao L, Wang X, Xu Y, Zhang Q (2011) Spectrum trading in cognitive radio networks: a contract-theoretic modeling approach. IEEE J. Sel. Areas Commun. 29(4):843–855

Han Z, Niyato D, Saad W, Başar T, Hjørungnes A (2011) Game theory in wireless and communication networks theory, models, and applications. Cambridge University Press, Cambridge, UK

Hasan Z, Bhargava V (2013) Relay selection for ofdm wireless systems under asymmetric information: a contract-theory based approach. IEEE Trans. Wirel. Commun. 12(8):3824–3837

Hossain E, Niyato D, Han Z (2009) Dynamic spectrum access and management in cognitive radio networks. Cambridge University Press, Cambridge, UK

Kim H, Shin KG (2008) Efficient discovery of spectrum opportunities with mac-layer sensing in cognitive radio networks. IEEE Trans Mobile Comput 7(5):533–545

Laffont JJ, Tirole J (1988) The dynamics of incentive contracts. Econometrica 56(5):1153–1175

Letaief K, Zhang W (2009) Cooperative communications for cognitive radio networks. Proc IEEE 97(5):878–893

Nie N, Comaniciu C (2005) Adaptive channel allocation spectrum etiquette for cognitive radio networks. In: First IEEE international symposium on new frontiers in dynamic spectrum access networks (DySPAN), pp 269–278

Niyato D, Hossain E, Han Z (2009) Dynamics of multiple-seller and multiple-buyer spectrum trading in cognitive radio networks: a game-theoretic modeling approach. IEEE Trans Mobile Comput 8(8):1009–1022

Pan M, Li P, Song Y, Fang Y, Lin P (2012) Spectrum clouds: a session based spectrum trading system for multi-hop cognitive radio networks. In: Proceedings of IEEE conference on computer communications, INFOCOM 2012, Orlando, FL

Roland G (2000) Transition and economics. The MIT Press, Cambridge, MA

Scott WR (2014) Financial accounting theory. Pearson Education Canada

Wang S, Xu P, Xu X, Tang S, Li X, Liu X (2010a) Toda truthful online double auction for spectrum allocation in wireless networks. In: IEEE Symposium on new frontiers in dynamic spectrum, Singapore

Wang X, Li Z, Xu P, Xu Y, Gao X, Chen HH (2010b) Spectrum sharing in cognitive radio networks: an auction-based approach. IEEE Trans. Syst. Man Cybernet. Part B (Cybernetics) 40(3):587–596

Zhu X, LS, Yum TSP (2007) Analysis of cognitive radio spectrum access with optimal channel reservation. IEEE Commun Lett 11(4):304–306

Xie R, Yu FR, Ji H (2012) Spectrum sharing and resource allocation for energy-efficient heterogeneous cognitive radio networks with femtocells. In: IEEE international conference on communications (ICC), Ottawa, Canada, pp 1661–1665

Zhang Y, Gu Y, Song L, Dawy Z, Han Z (2015a) Tournament based incentive mechanism designs for mobile crowdsourcing. In: IEEE global communications conference (GLOBECOM), San Deigo, CA

Zhang Y, Song L, Saad W, Dawy Z, Han Z (2015b) Contract-based incentive mechanisms for device-to-device communications in cellular networks. IEEE J. Sel. Areas Commun. (JSAC) 33(10):2144–2155

Zhou X, Zheng H (2009) Trust: a general framework for truthful double spectrum auctions. INFOCOM 2009. IEEE, Rio de Janeiro, Brazil, pp 999–1007

Chapter 7
Complementary Investment of Infrastructure and Service Providers in Wireless Network Virtualization

7.1 Introduction

The contract theoretical models adopted previously are all belong to the complete contract. While in wireless networks, there are also problems that are in need of designing incomplete contracts to solve. We will give one example of incomplete contract in virtualized network, which is about relationship-specific investment. The rapid evolution of information and communication technologies and infrastructure has been a key motivator for reducing the cost of wireless network deployment and operation. The premise of creating "virtual" resources, such as infrastructure and spectrum resources, that can be shared led to the emergence of the notion of wireless network virtualization (Liang and Yu 2015b). One widely adopted mobile virtual network (MVN) framework is the two-level business model shown in Fig. 7.1. In this model, infrastructure providers (InPs) deliver physical wireless network resources, such as towers, base stations, and radio spectrum. Meanwhile, service providers (SPs) act as mobile virtual network operators (MVNOs) who operate, program, and lease the virtual resources while also offering end-to-end service to end users (Liang and Yu 2015a).

For both InPs and SPs, the main investments include capital expenditure (CapEx) and operational expenditure (OpEx), which are used to implement the network infrastructure and support its operation (Bari et al. 2013). CapEx is the prominent investment of an InP, and it includes the cost of purchasing and installing equipment such as base stations, backhaul aggregators, radio network controller, core network (CN), and the cost of using licensed spectrum issued by the authorities (Celentano 2015). An InP's OpEx includes energy charge, human resources that are employed in site and backhaul lease, operation, and maintenance. Similarly, an SP will incur CapEx and OpEx when executing the virtualization process, initializing and maintaining the end-to-end services for end users (Chase and Niyato 2015).

In real scenarios, the InPs and SPs usually sign a long-term supply contract on a base price and subject to price adjustment according to the future market. Indeed, in an MVN, the InP and SP must work together to reap the benefits of their investments.

© Springer International Publishing AG 2017
Y. Zhang and Z. Han, *Contract Theory for Wireless Networks*,
Wireless Networks, DOI 10.1007/978-3-319-53288-2_7

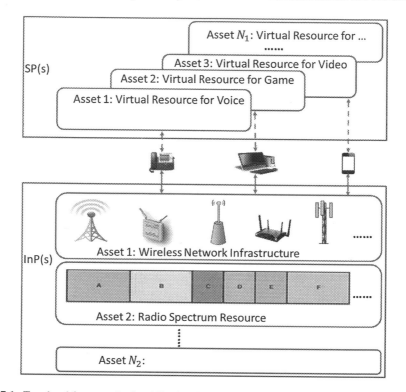

Fig. 7.1 Two-level framework of mobile virtual networks

On the one hand, the InP provides the physical wireless network that enables the SP to serve end users. On the other hand, the SP pays the InP for providing the platform to transmit its data. Such a relationship between InP and SP will be referred to as a *complementary relationship* hereinafter.

The number of users attracted by SPs and the amount of mobile data traffic served by InPs are the two main factors that drive the expansion of the MVN concept. However, any upgrade and expansion of MVNs will require further investments. For example, to increase the coverage and capacity of the physical wireless network, the InP must acquire more bandwidth and more capable equipment, which increases both CapEx and OpEx. If the SP plans to expand its market and attract more users, investment in human capital may be needed as well to develop new online services.

Due to the complementary relation between SP and InP, their bargaining positions and future returns depend not only on the market, but also on the investments they have made ex ante. For example, seeing that there is a growing market of LTE, one InP plans to spend vast sums in customizing its radio access network to fit the special needs of an SP who is a telecom carrier. This customized investment will increase the InP's efficiency in trading with this specific SP. However, the radio access technologies in wireless networks are different and often incompatible between operators. Thus, the InP will reduce the opportunities that it can create with other SPs, ex post. In contrast, if the InP anticipates such a weak negotiation position, it may refrain from making such SP-specific investments even if they are efficient. Similar decisions are hard to make for the SP to initiate new online services for users.

Clearly, to achieve high efficiency of an MVN, the ex ante investments in SP and InP are critical. To solve this network efficiency problem, we need to answer the questions such as when the physical and virtual resources are owned by different parties, how much should the InP and SP invest in the network expansion? If the ownerships of the key resources in MVN can be integrated into one, what is the optimal investment policy? The main contribution of this chapter is to study the problem of how the ownerships of resources affect the investment efficiency in an MVN and answer the raised questions. The trading between InP and SP shows the property of a complementary relation. The developed model is generic enough to accommodate multiple SPs and InPs, as well as multiple physical and virtual resources. Subsequently, for the special case in which there are only single SP and InP, we provide a detailed analysis of cases in which the physical and virtual resources are owned separately or integrated. Last but not least, we investigate the parameters that influence the efficiency of investment through simulations.

The rest of this chapter is organized as follows. First, we will introduce the mobile virtual network complementary investment model in Sect. 7.2. Then, problem formulation for the general case is described in Sect. 7.3, in which we place the emphasis on the analysis of the special case where the number of physical resources, virtual resources, InPs, and SPs in the MVN are all equal to one. The performance evaluation is conducted in Sect. 7.4. Finally, summary is given in Sect. 7.5.

7.2 System Model

Consider an MVN composed of a set of InPs represented by $\mathcal{J}$ and a set of SPs denoted as $\mathcal{K}$. The InPs own multiple physical resources such as the licensed spectrum, sites (towers and antennas), base stations (macrocell, small cell), access points, CN elements (gateway, switchers, routers). The virtual resources owned by SPs include all the virtual entities sliced by each element in the physical wireless network. This model is aligned with the field of property ownership theory discussed in Hart and Moore (1990).

All InPs and SPs are assumed to be risk neutral and pool together as a set of I agents, $\mathcal{S} = \mathcal{J} \cup \mathcal{K}$, and each agent is denoted by $i = 1, \ldots, I$. Any subset of agents is denoted as $S \subseteq \mathcal{S}$. Furthermore, the set of all physical and virtual resources that are available in the MVN is denoted by $\mathcal{A}$ with N resources $(a_1, a_2, \ldots, a_N)$, with the subset of resources $A \subseteq \mathcal{A}$.

As previously discussed, we model the InPs and SPs in the MVN as being engaged in a complementary relation. The InPs and SPs' investments include expenditures in capital and human resources that are more or less specific to the resources in $\mathcal{A}$ and thus affect the InPs and SPs' productivity and bargaining position in the future. An agent can choose what type of investment to make (or type of service to provide). As a simplification, we restrict our attention to the case in which the types of services offered by the InPs and SPs are fixed; they choose only what level of service to provide. For example, InPs invest to expand their network capacity and coverage, which can include wider spectrum bandwidth and larger antenna gain.

For expanding an MVN, the investments of the InPs and SPs can be viewed as a two-stage problem. In the first stage, each agent i makes ex ante investment x_i on its resources at a cost $\psi_i(x_i)$. Then, in the second stage, the trade among a subset of InPs and SPs $S \subseteq \mathscr{S}$ combined with the subset of resources $A \subseteq \mathscr{A}$ begins, and a revenue $V(S, A, \mathbf{x})$ is generated, where $\mathbf{x} = (x_1, x_2, \ldots, x_I)$ denotes the vector of all InPs and SPs' ex ante investments. Since each agent only chooses how much to invest, we suppose that x_i is a scalar in the range $[0, \bar{x}_i]$.

7.2.1 Cost and Revenue Functions

7.2.1.1 Cost

The InPs and SPs will incur monetary costs when making such investments. For different types of investments, the cost functions $\psi_i(x_i)$ can differ. Here, we assume that the cost function ψ_i is twice differentiable and strictly increasing and strictly convex with respect to the investment x_i, i.e., $\psi_i'(x_i) \geq 0$ and $\psi_i(x_i) = 0$. If $x_i > 0$, then $\psi_i'(x_i) > 0$ and $\psi_i''(x_i) > 0$ for $x_i \in (0, \bar{x}_i)$, with $\lim_{x_i \to 0} \psi_i'(x_i) = 0$ and $\lim_{x_i \to \bar{x}_i} \psi_i'(x_i) = \infty$.

7.2.1.2 Revenue

Consider a *coalition* of InPs and SPs in a subset of S who control a subset of resources A. The revenue $V(S, A, \mathbf{x})$ obtained from the trading within this coalition is also twice differentiable, strictly increasing, and measured in monetary terms. But $V(S, A, \mathbf{x})$ is concave in x_i instead of convex as the cost functions. There are two constraints on $V(S, A, \mathbf{x})$

$$\frac{\partial V(S, A, \mathbf{x})}{\partial x_i} = 0, \quad if \ i \notin S, \tag{7.1}$$

$$\frac{\partial^2 V(S, A, \mathbf{x})}{\partial x_i \partial x_j} \geq 0, \quad \forall j \neq i, \tag{7.2}$$

where (7.1) implies that the InP and SP's marginal investment affects only the value of coalitions of which it is a member and (7.2) denotes the complementary relation of the investment, that is, if an InP invests to upgrade its physical wireless network capacity, the SPs can also benefit from that.

7.2.2 Shapley Value

In this general setup with I InPs and SPs in a MVN, the main difficulty is to nego-
tiate how the revenue of the trade gets determined, in other words how the ex post
revenue $V(S, A, \mathbf{x})$ gets divided up among the InPs and SPs in the coalition. In
Shapley (1953), the proposed solution is to assume that the outcome of multilateral
negotiations is divided according to the *Shapley value*.

When a coalition of InPs and SPs S decides to form an MVN, they agree to pool all
the physical and virtual resources owned by any of the members. Then, the mapping
$\omega(S)$ from $\mathscr{S}$ to $\mathscr{A}$ denotes the subset of resources owned by the subset of agents
S (Grossman and Hart 1986). As done in Bolton and Whinston (1993), we assumed
that each resource can be controlled by at most one of the coalitions of agents S, or
its complement $\mathscr{S} \backslash S$. In addition, we assume that the resources controlled by some
subset $S' \subseteq S$ must also be controlled by the whole coalition S. Thus, we have the
following properties for the mapping $\omega(S)$:

$$\omega(S) \cap \omega(\mathscr{S} \setminus S) = \emptyset, \tag{7.3}$$

$$\omega(S') \subseteq \omega(S), \tag{7.4}$$

$$\omega(\emptyset) = \emptyset. \tag{7.5}$$

The *Shapley value* assigns a revenue to an agent i possibly involved in a transaction
with S agents who together own or control $\omega(S)$ resources. We give the formal
definition of *Shapley value* as follows:

Definition 7.1 Given an ownership allocation $\omega(S)$, a vector of ex ante investments x, and the associated ex post revenue for any given coalition of agents S,
$V(S, \omega(S), \mathbf{x})$, the *Shapley value* specifies the following expected ex post revenue
for any agent i:

$$B_i(\omega, \mathbf{x}) = \sum_{S|i \in S} p(S)[V(S, \omega(S), \mathbf{x}) - V(S \setminus i, \omega(S \setminus i), \mathbf{x})], \tag{7.6}$$

where
$$p(S) = \frac{(s - 1)!(I - s)!}{I!}, \tag{7.7}$$

and $s = |S|$ is the number of agents in S.

To summarize, the *Shapley value* is an expected revenue, where the expectations
are taken over all possible subcoalitions S that agent i might join ex post. That is,
each agent looks at ex post coalition formation like a random process where any order
in which coalitions get formed is equally likely (Bolton and Dewatripont 2004). It
is for this reason that the probability distribution $p(S)$ is as specified in (7.7). Given
any ex post realization of a coalition, S, the *Shapley value* assigns to each agent i

in the coalition the difference in surplus obtained with the entire group S and with the coalition excluding agent i. In other words, the *Shapley value* assigns to each agent i the expected contribution of that agent to the overall ex post revenue obtained through multilateral trade between all agents.

7.2.3   Investment Surplus

Given the *Shapley value* as the expected revenue and the cost of InP and SP's investments, we have the surplus of each agent's investment:

$$R_i(S, \omega(S), \mathbf{x}) = B_i(\omega, \mathbf{x}) - \psi_i(x_i). \tag{7.8}$$

7.3   Problem Formulation

Given the InPs and SPs' revenue and cost functions of investment in MVN expansion, we are about to see what is the optimal investment level when they are chosen non-cooperatively by formulating the InPs and SPs' surplus maximization problem in this section. We will first give the general results where there are multiple resources, as well as multiple InPs and SPs. Next, we will give the representative case, where there are only one InP and SP, with only one physical resource and virtual resource. In this example, we will give additional analysis on how the ownerships of those resources affect the investment incentives of the InP and SP in a MVN.

7.3.1   General Case

Within one coalition S of multiple InPs and SPs, member chooses the investment x_i non-cooperatively to maximize their respective expected surplus:

$$\max_{x_i} B_i(\omega, \mathbf{x}) - \psi_i(x_i). \tag{7.9}$$

The optimal investment x_i^* is characterized by the first-order conditions of (7.9):

$$\frac{\partial B_i(\omega, \mathbf{x})}{\partial x_i} = \sum_{S|i \in S} p(S) \frac{\partial V(S, \omega(S), \mathbf{x})}{\partial x_i}, \tag{7.10}$$

$$= \psi_i'(x_i).$$

Different coalitions of InPs and SPs will result in different optimal investment levels, because different ownerships of the physical and virtual resources affect the incentives of the InPs and SPs in investment. Next, we are going to see how the optimal investment levels are affected when the physical resource and virtual resource are owned separately or together by InP and SP.

7.3.2   Single Provider and Single Resource

Consider a two-level business model for MVNs in which we have only one SP operating on one virtual resource and a single InP working on one physical resource. Thus, we have $I = 2$ and $\mathbb{A} = \{a_1, a_2\}$ and denote the InP as agent 1 and the SP as agent 2. Each agent can make ex ante investments x_i in a first stage, and the trade between the InP and SP takes place in a second stage.

In general, we can have three different scenarios: non-integration, InP integration, and SP integration (Bolton and Whinston 1993). Non-integration means the ownerships of the physical and virtual resources are separated, physical resource is under the control of InP, and virtual resource is controlled by SP. InP integration means that the InP is the owner of both physical and virtual resources and SP can only operate and use the virtual resource under the allowance of the InP. In contrast, the SP has the ownership of both physical and virtual resources under the SP integration, and InP has limited access to the physical resource.

Based on this interpretation, we can set up the system model as follows:

Non-integration: $\omega(1) = \{a_1\}, \omega(2) = \{a_2\}$;

InP integration: $\omega(1) = \emptyset, \omega(2) = \{a_1, a_2\}$;

SP integration: $\omega(1) = \{a_1, a_2\}, \omega(2) = \emptyset$.

7.3.3   Non-integration

Due to the complementary relation between InP and SP, it is intuitive to see that no ex post revenue can be generated without combining the physical and virtual resources together in an MVN. Then, under non-integration, the ex post revenue that is generated by a single InP or SP is as follows:

$$V(\{1\}, a_1, \mathbf{x}) = V(\{2\}, a_2, \mathbf{x}) = 0, \tag{7.11}$$

where $x = (x_1, x_2)$.

If, however, both InP and SP form a coalition by trading access to their respective resources, they generate a strictly positive revenue:

$$V(\{1, 2\}, \{a_1, a_2\}, \mathbf{x}) = V(\mathbf{x}) > 0, \tag{7.12}$$

where $V(\mathbf{x})$ is the maximum revenue obtain by $V(S, \omega(S), \mathbf{x})$. Since there are only two equally likely orderings of coalition formation, $\{1, 2\}$ and $\{2, 1\}$, we have $p(\{1, 2\}) = p(\{2, 1\}) = \frac{1}{2}$. Under non-integration, the *Shapley value* then assigns an expected revenue to the InP and SP as

$$B_1(NI|\mathbf{x}) = B_2(NI|\mathbf{x}) = \frac{1}{2}V(\mathbf{x}), \tag{7.13}$$

where NI stands for non-integration.

Based on our assumptions that $V(S, \omega(S), \mathbf{x})$ is strictly increasing and concave in $x = (x_1, x_2)$ and the investment cost functions $\psi_i(x_i)$ are strictly increasing and convex in x_i, the InP and SP choose their ex ante investment non-cooperatively to maximize their respective expected revenue:

$$\max_{x_i} \frac{1}{2}V(\mathbf{x}) - \psi_i(x_i). \tag{7.14}$$

Due to the concavity of the objective function, these equilibrium investment levels can be obtained from the first-order conditions of each party's optimization problem:

$$\frac{1}{2}\frac{\partial V(x_1, x_2)}{\partial x_i} = \psi_i'(x_i). \tag{7.15}$$

Thus, under non-integration, the equilibrium investment levels (x_1^{NI}, x_2^{NI}) are given by

$$\frac{1}{2}\frac{V(x_1^{NI}, x_2^{NI})}{\partial x_1} = \psi_1'(x_1^{NI}), \tag{7.16}$$

$$\frac{1}{2}\frac{V(x_1^{NI}, x_2^{NI})}{\partial x_2} = \psi_2'(x_2^{NI}). \tag{7.17}$$

7.3.4 Infrastructure Integration

Under InP integration, the InP owns both the physical and virtual resources in the MVN, and the SP can only operate the virtual resource with the InP's permission. Then, it is possible for the InP to generate an ex post revenue on its own, since the resources it owns are sufficient to run as a complete MVN. Nevertheless, the SP cannot generate any revenue on its own, as under non-integration. So we still have (7.12), as well as

$$V(\{2\}, \emptyset, \mathbf{x}) = 0. \tag{7.18}$$

When the InP operates both physical and virtual resources, the ex post revenue that can be generated with only InP is as follows:

$$V(\{1\}, a_1, a_2, \mathbf{x}) = \Phi_1(x_1), \tag{7.19}$$

where $\Phi_1(x_1)$ is the InP's revenue function obtained from $V(S, \omega(S), \mathbf{x})$. Due to the complementary relation with SP, it is plausible that the InP might be able to make higher revenue by hiring the SP to operate the virtual resource. Thus, the revenue InP obtains by itself is lower than the case when it cooperates with SP, i.e., $\Phi_1(x_1) < V(\mathbf{x})$.

The *Shapley value* under InP integration is then given by

$$B_1(InPI|\mathbf{x}) = \frac{1}{2}[V(\mathbf{x}) - \Phi_1(x_1)] + \Phi_1(x_1), \tag{7.20}$$

$$B_2(InPI|\mathbf{x}) = \frac{1}{2}[V(\mathbf{x}) - \Phi_1(x_1)]. \tag{7.21}$$

Thus, under InP integration, the equilibrium investments (x_1^{InPI}, x_2^{InPI}) are given by

$$\frac{1}{2}\frac{V(x_1^{InPI}, x_2^{InPI})}{\partial x_1} + \frac{1}{2}\Phi_1'(x_1^{InPI}) = \psi_1'(x_1^{InPI}), \tag{7.22}$$

$$\frac{1}{2}\frac{V(x_1^{InPI}, x_2^{InPI})}{\partial x_2} = \psi_2'(x_2^{InPI}). \tag{7.23}$$

7.3.5   Service Provider Integration

Indeed, the SP integration is the mirror image of InP integration, so that the *Shapley value* under SP integration becomes

$$B_1(SPI|\mathbf{x}) = \frac{1}{2}[V(\mathbf{x}) - \Phi_2(x_2)], \tag{7.24}$$

$$B_2(SPI|\mathbf{x}) = \frac{1}{2}[V(\mathbf{x}) - \Phi_2(x_2)] + \Phi_2(x_2), \tag{7.25}$$

where $\Phi_2(x_2)$ is the SP's revenue function obtained from $V(S, \omega(S), \mathbf{x})$ and is also lower than the case when it cooperates with InP. Thus, under SP integration, the equilibrium investment levels (x_1^{SPI}, x_2^{SPI}) are given by

$$\frac{1}{2}\frac{V(x_1^{SPI}, x_2^{SPI})}{\partial x_1} = \psi_1'(x_1^{SPI}), \tag{7.26}$$

$$\frac{1}{2}\frac{V(x_1^{SPI}, x_2^{SPI})}{\partial x_2} + \frac{1}{2}\Phi_2'(x_2^{SPI}) = \psi_2'(x_2^{SPI}). \tag{7.27}$$

7.3.6 Summary

When the InP and SP are not integrated, any party can make the ex ante investments according to the equilibrium obtained from the optimization problem. If the ownership of both resources is integrated, for example, under InP integration where the InP is the sole owner of both the physical and virtual resources, then its ex post-negotiating position with the SP is less affected by specific investments. The InP would of course be inclined to make any ex ante-specific investments that are efficient. But as the InP is the sole owner of both resources, the SP is now the InP's employee, and the SP would have less incentive to invest than under non-integration.

In summary, the ownership allocation affects the InP and SP's incentives in specific investments. If investments in customized infrastructures are most valuable, then it makes sense for the InP to own physical resources and the SP's business. If investments in virtual resource operation and end-to-end service are most valuable, then it makes sense for the SP to own the virtual resource and the physical resource. Finally, if both types of investment are important, it may be best to separate their business.

7.4 Simulation Results and Analysis

In this section, we will give numerical simulations to illustrate how the InP and SP's incentives to invest are affected by the ownership of resources. First, we will give the specific form of the revenue and cost functions we have defined in the system model. Then, we will show the InP and SP's optimal investment level and surplus by varying the cost coefficient and marginal return and do a comparison between InP and SP when resources are under different ownerships, i.e., non-integration, InP integration, and SP integration.

7.4.1 Simulation Setup

In the system model, we have defined the revenue function V as a concave function. Here, we choose a logarithmic function for the revenue V as follows:

$$V(\mathbf{x}) = \log_n(1 + \sum_{i=1}^{N} x_i).\tag{7.28}$$

The InP and SP's solo revenue function under integration is

$$\Phi_i(x_i) = \log_n(1 + x_i).\tag{7.29}$$

Clearly, $\Phi_i(x_i) < V(x)$ is satisfied. The partial derivative $\Phi_i'(x_i)$ is the marginal return of each investment. By varying the index n, we can change the marginal return of different investments.

Furthermore, we have defined the cost function in the system model as a convex function. Here, we set up the cost function ψ_i in a quadratic form as

$$\psi_i(x_i) = \frac{1}{2}a_i x_i^2, \tag{7.30}$$

where a_i is the cost coefficient of each investment. From the previous section, we see that the SP integration is the mirror image of the InP integration. In order to distinguish the investment and surplus of InP and SP, we assign a higher cost coefficient a_1 to InP than that of SP (a_2).

7.4.2 Cost Coefficient

In Figs. 7.2 and 7.3, we study the cost coefficient's impact on the optimal investment level and surplus and do comparisons between InP and SP under different ownership scenarios. From the simulation results, we can see that as the magnification of the cost coefficient a_i increases, the investment and surplus also decrease. The reason for this phenomenon is that a larger cost coefficient a_i means more cost when making an investment. In such a case, both InP and SP are less likely to invest in the MVN. With less investment, the network capacity will decrease, and the InP and SP's surplus will

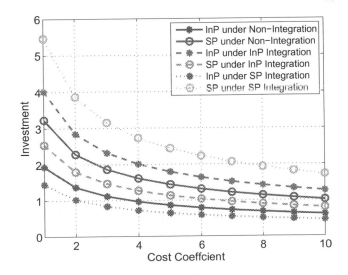

Fig. 7.2 The impact of the cost coefficient on investment

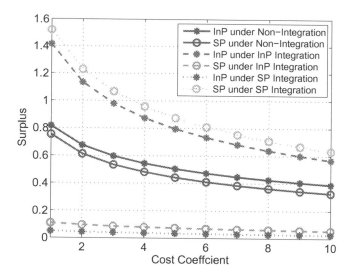

Fig. 7.3 The impact of the cost coefficient on surplus

certainly decrease. Furthermore, from Fig. 7.2, we see that the InP invests more in physical resources under InP integration, and the SP invests more in virtual resources under SP integration. In contrast, the physical resource receives the least investment under SP integration, and virtual resource receives the least investment under InP integration.

7.4.3 Marginal Return

In Figs. 7.4 and 7.5, we study how the marginal return affects the InP and SP's investment and surplus while fixing the cost coefficient. From Fig. 7.4, we can see that as the index n increases, the investments increase under integration, but decrease under non-integration. The reason is that when $0 < \Phi_i'(x_i) < \partial V(x)/\partial x_i$, integration always induces higher incentives for the InP and SP than non-integration. From Fig. 7.5, we can see that both InP and SP result in a decrease of surplus when marginal return increases. This result is due to that when the marginal return $\Phi_i'(x_i)$ is large, either form of integration will result in overinvestment. In the case of InP integration and SP integration, the overinvestments result in undesirable surpluses for SP and InP, respectively. Similar to the previous result, we see that the InPs have less incentive to invest under non-integration or under SP integration than when itself owns the integrated MVN. The same is for the SP.

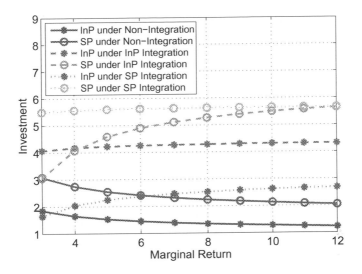

Fig. 7.4 The impact of the marginal return on investment

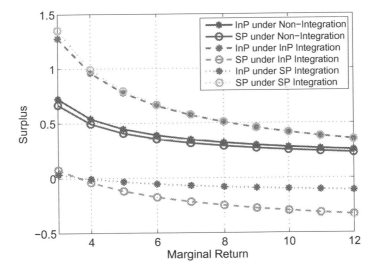

Fig. 7.5 The impact of the marginal return on surplus

7.5 Summary

In this chapter, we have given one application of incomplete contract in relationship-specific investment by considering the problem of how to efficiently make investments to expand MVN capacity and coverage under the complementary relation between InPs and SPs. We not only gave the general model and solution when there are multiple physical and virtual resources in the MVN and the InPs and SPs that own and operate them. We but also analyzed the problem of how the ownership of physical and virtual resources affects the InP and SP's incentives in investment, especially in the case when they are owned separately or integrated. Using simulations, we have shown that the ownership and marginal return of resources affect InP and SP's incentives to invest and surpluses.

References

Bari M, Boutaba R, Esteves R, Granville L, Podlesny M, Rabbani M, Zhang Q, Zhani M (2013) Data center network virtualization: a survey. IEEE Commun Surv Tutorials 15(2):909–928
Bolton P, Dewatripont M (2004) Contract theory. The MIT Press, Cambridge, MA
Bolton P, Whinston MD (1993) Incomplete contracts, vertical integration, and supply assurance. Rev Econ Stud 60(1):121–148
Celentano J (2015) U.S. wireless CapEx looking up. http://www.aglmediagroup.com/u-s-wireless-capital-expenditures-looking-up/
Chase J, Niyato D (2015) Joint optimization of resource provisioning in cloud computing. IEEE Trans Serv Comput (99):1–1
Grossman SJ, Hart OD (1986) The costs and benefits of ownership: a theory of vertical and lateral integration. J Polit Econ 94(4):691–719
Hart O, Moore J (1990) Property rights and the nature of the firm. J Polit Econ 98(6):1119–1158
Liang C, Yu F (2015a) Wireless network virtualization: a survey, some research issues and challenges. IEEE Commun Surv Tutorials 17(1):358–380
Liang C, Yu F (2015b) Wireless virtualization for next generation mobile cellular networks. IEEE Wireless Commun 22(1):61–69
Shapley LS (1953) Stochastic games. Proc Nat Acad Sci USA 39(10):1095–1100

Chapter 8
Conclusion and Future Works

8.1 Conclusion Remarks

In this book, we have provided the contract theory framework for wireless networking, which has been sequentially awarded with the Nobel Prize in economics science for 2014 and 2016. Contract theory is highly evaluated due to its effectiveness in market power and regulation—specifically how to regulate oligopolies in situations with asymmetric information, i.e., when regulators do not know everything about how firms are operating. Meanwhile, contract theory itself is an efficient tool in dealing with asymmetric information between employer/seller(s) and employee/buyer(s) by introducing cooperation. Such a framework for designing regulations has been applied to a number of industries, from banking to telecommunications. Given the properties of wireless networks, which encounter many situations of asymmetric information and the need for cooperation, contract theory is an excellent tool by modeling the employer/seller(s) and employee/buyer(s) as different roles depending on the scenario under consideration.

This book provides a theoretical research between wireless communications, networking, and economics, in which different contract theory models have been applied in various wireless networks scenarios. We start with the fundamental concepts of contract theory and introduced the potential applications for each class of the typical contract problems: *adverse selection*, *moral hazard*, and the mixed of them two. Specially, we have investigated the design of reward, which is the most critical element in an incentive mechanism design. We have also provided a detailed description of the potential of using such contract-theoretic tools in several wireless applications, such as spectrum trading cognitive radio network, relay selection, distributed computing, D2D communication, and mobile crowdsourcing.

In the first application, the problem of pure *adverse selection* is studied to solve the incentive problem of encouraging cellular UEs to participate in D2D communication underlaid cellular network. Given the information asymmetry that the UEs' preferences are unobservable to the BS, we have proposed a self-revealing mechanism that forces UEs to select the contracts that are in consistent with their preferences.

© Springer International Publishing AG 2017
Y. Zhang and Z. Han, *Contract Theory for Wireless Networks*,
Wireless Networks, DOI 10.1007/978-3-319-53288-2_8

Simulation results have shown that the proposed approach outperforms the linear pricing which does not try to retrieve any information at all, but cannot compete with the optimal contract with no information asymmetry.

Next, the problem of pure *moral hazard* is studied to investigate the issue of providing incentives for smart device users to participate in mobile crowdsourcing and designing optimal payment plan for fog nodes to cooperate in fog computing. In particular, not only we have solved the problem in the basic one-dimensional and one-user case, but also we extend the model into multi-dimensional and multi-user cases. The basic one-dimensional and one-user model is first applied in mobile crowdsourcing, in which one user is awarded by the principal's evaluation from a single aspect. Then, we extended the case into the multi-user one, where the principal rewards users based on the rank of their performance as in a tournament. The optimal contract is solved as a fixed list of prizes. The optimal contract is solved as a bundle of reward and effort. The last application of *moral hazard* is in fog computing, in which a multi-dimensional model is applied to solve the optimal payment plan while ensuring the FN's cooperation. In the three applications, the numerical results showed the comparisons between the utilities in the optimal contracts and other different incentive mechanisms and analyzed that the principal's utility varies with different parameters such as operation cost coefficient, risk aversion degree, and measurement error variance.

Furthermore, the mixed problem of both the *adverse selection* and *moral hazard* problem is studied to address the problem of spectrum trading in a cognitive radio network. The unobservable of SU's capability in generating revenue from utilizing the spectrum is modeled as *adverse selection* and the unobservable of SU's effort putting into utilizing the spectrum is modeled as *moral hazard*. The three different problems, i.e., two extreme cases where only *adverse selection* or *moral hazard* is present and the general case where both are present, are solved and analyzed. Through extensive simulations, we have also shown different parameters' effects on the system performance and showed that the two extreme cases serve as the upper and lower bound for the general case where both problems are present.

Finally, we make progress toward the incomplete contract by studying the problem of how to efficiently make investments to expand MVN capacity and coverage under the complementary relation of InPs and SPs. We not only gave the general model and solution when there are multiple physical and virtual resources in the MVN and the InPs and SPs that own and operate them. We also analyzed the problem of how the ownership of physical and virtual resources affects the InP and SP's incentives in investment, especially in the case when they are owned separately or integrated. Using simulations, we have shown that the ownership and marginal return of resources affect InP and SP's incentives to invest and surpluses.

From those works, we have seen contract theory as a useful framework to design incentive mechanisms to motivate the third party's cooperation in emerging wireless networks, such as heterogeneous networks, D2D communication, mobile crowdsouring, mobile cloud computing, and cognitive radio networks. In a nutshell, this book is expected to provide an accessible and holistic survey on the use of new techniques from contract theory to address the future of network economics area, and have a long-term effect on problems such as incentive mechanism and pricing schemes design, resource sharing and trading.

8.2 Future Work

Under the background of rapid development wireless networks and the proliferation of highly capable mobile devices, cooperations in wireless networks are and will be in highly demand in numerous areas. Incentive mechanism design is to ensure that cooperation falls into the emerging world-class high-impact theoretical research between wireless communications, networking, and economics. Thus, we see there is a huge potential to do further research in incentive mechanism design and use contract theory to solve cooperation problems in wireless networks. The following is research directions that can be further explored in this area of research.

- *Exploring emerging wireless network applications*: There are many areas in wireless networks where the cooperation among different parties is extremely needed. Some interesting areas where cooperation is called upon to play a key role include wireless network virtualization, cloud radio access networks, physical layer security, multimedia distribution in ultra-dense networks, and load management in wireless networks with machine-to-machine communications.
- *Exploring new contract theory models*: First, current applications in wireless networks do all belong to the static basic and extended models in multi-dimensional and multi-lateral *adverse selection* and *moral hazard* models. In the future works, we can extend the static models into repeated contracting and incomplete contracting ones, which show great potential in modeling more sophisticated interplay between different parties. Second, contract theory can be utilized to address wireless networking problems other than cooperation incentives. There are other models in contract theory that provide potential techniques, e.g., utilizing insurance design and audition in mobile cloud computing or using system hierarchy efficiency in infrastructure deployment.
- *Exploring the connection between wireless physical meanings and economic factors*: By applying this microeconomic model into wireless networks, it is important to well model and define the economic parameters with appropriate wireless communication network physical meanings, since the ultimate goal of using contract theory here is to respond to the technical problems in wireless networks. Without properly characterizing the wireless network system, the solution will be less meaningful and infeasible to apply.

Printed in the United States
By Bookmasters